EXPERT SYSTEMS

IN CIVIL ENGINEERING

Proceedings of a Symposium sponsored by the
Technical Council on Computer Practices
of the American Society of Civil Engineers
in conjunction with the ASCE Convention in
Seattle, Washington
April 8-9, 1986

**Edited by
Celal N. Kostem
and Mary Lou Maher**

Published by the
American Society of Civil Engineers
345 East 47th Street
New York, New York 10017-2398

The Society is not responsible for any statements
made or opinions expressed in its publications.

FOREWORD

This volume contains the papers accepted for the "First Symposium on Expert Systems in Civil Engineering," held in Seattle, Washington, April 8–9, 1986. The symposium is sponsored by the Technical Council on Computer Practices, ASCE.

Over the years a number of sessions have been conducted at ASCE conventions with the theme of computer-aided engineering. The First Symposium on Computer Aided Design in Civil Engineering, sponsored by the Technical Council on Computer Practices, was held in San Francisco, California, in October, 1984. The sessions, and especially the 1984 symposium indicated the growing importance of expert systems, also referred to as knowledge based expert systems, in civil engineering practice. It was thus decided to have a symposium dedicated only to expert systems.

Some of the authors of the papers herein were requested to elaborate on the definitions and characteristics of expert systems. This was done to permit the proceedings to be used as a teaching and learning tool. The authors kindly cooperated, thus in addition to covering extensive research and application areas, they also provided the needed introductions to expert systems. The symposium organizers extend sincere thanks to all authors for their cooperation.

All papers are eligible for discussion in the Journal of Computing in Civil Engineering. All papers are eligible for ASCE awards.

The editors are grateful to Ms. Shiela Menaker, Manager, Book Production Department, ASCE for her continual work and guidance in the preparation of the proceedings, and especially for her patience in the efforts to meet various deadlines.

Celal N. Kostem
Mary Lou Maher
Editors, The First Symposium
on Expert Systems in
Civil Engineering

CONTENTS

*Manuscript not available at time of printing

WHAT IS AN EXPERT SYSTEM

Steven J. Fenves[1]

Member, ASCE

Introduction

Knowledge-based expert systems (KBES) have created as much excitement in the civil engineering computer user community as the emergence of FORTRAN in the '50's, problem-oriented languages in the '60's, and CAD in the '70's. On the one hand, KBES evoke expectations of full-blown Artificial Intelligence (AI) systems: programs that adapt, learn, invent, and accumulate the combined wisdom of a profession. On the other hand, many application programmers find themselves in the position of the character M. Jourdain, in Moliere's play, *Les Femmes Savantes*, who gleefully discovers that he "has been writing prose all his life".

This paper is presented in an attempt to define and clarify the definition, role and impact of KBES in civil engineering. It will be shown that most expectations of AI are largely premature. The second view, that of Moliere's character, is more difficult to address. Certainly, many conventional algorithmic design application programs have always incorporated expertise in the form of limitations, assumptions, approximations, shortcuts, etc. The paper contrasts these programs to KBES in a number of dimensions, such as control, knowledge structure, and development process. However, it will be shown that the fictional character's attitude will eventually prevail: the useful and practical KBES will be written by today's application programmers, once the KBES methodology is adopted as a component of their kit of tools.

Origins of KBES

In order to put the discussion in context, the origins of KBES need to be briefly described. Artificial Intelligence (AI) evolved as a branch of computer science, paralleling other branches of computer science such as languages, data structures, operating systems and numerical algorithms. It is worth remembering that the civil engineering computer user community has significantly benefited from the latter branches of computer science research, in the form of improved algorithmic languages, rugged algorithms, software engineering methods of tools, database methods, etc., which have vastly improved productivity in both program development and use.

The early AI efforts dealt with representation and processing of symbols (other than numbers) and emulation of information processing, including learning, of humans. The early AI problem-solvers solved tasks ranging from puzzle solving to theorem proving using only a minimal set of general problem-solving tools, such as heuristic search, hill climbing and means-end analysis; all domain-dependent

---

[1] University Professor of Civil Engineering, Carnegie-Mellon University, Pittsburgh, PA 15213.

1

knowledge was "learned" by these programs as they solved tasks. By the 1970's, it became clear to AI researchers that the above weak problem-solving methods were not sufficient. The first generation expert systems emerged when the problem-solving strategies were intimately combined with strong methods explicitly representing specific domain knowledge (e.g., infectious disease knowledge in MYCIN, geological knowledge in PROSPECTOR, molecular genetics knowledge in MOLGEN, etc.). The next major development, opening up the new field sometimes referred to as Knowledge Engineering, occurred when the problem-solving techniques and the domain-dependent knowledge were separated, permitting the combination of new domain knowledge with the existing problem-solving frameworks.

The following analogy, even if somewhat weak, may be helpful. In the 1950's, surveying and structural programs were written using only general tools, i.e., procedural languages. The original versions of COGO and STRESS combined these languages with domain-dependent procedural knowledge, providing vastly improved tools for surveyors and structural engineers, respectively, but not benefiting anyone else. The next development, ICES and its successor supervisory systems, permitted the combination of the general framework with any specialized set of procedures.

Standard Definition of KBES

A good standard definition of KBES is the following:

"Knowledge-based expert systems are interactive computer programs incorporating judgment, experience, rules of thumb, intuition, and other expertise to provide knowledgeable advice about a variety of tasks."[Gaschnig81]

The first reaction of many professionals active in computer-aided engineering to the above definition is one of boredom and impatience. After all, conventional computer programs for engineering applications have become increasingly interactive; they have always incorporated expertise in the form of limitations, assumptions and approximations; and their output has long ago been accepted as advice, not as "the answer" to the problem.

There is a need, therefore, to add an operational definition to distinguish the new wave of expert systems from conventional algorithmic programs which incorporate substantial amounts of domain-dependent heuristics. The distinction should not be based on implementation languages, e.g., FORTRAN vs. LISP (after all, EXPERT was written in FORTRAN, and several expert system development frameworks are now available in Pascal or C implementations), or any absolute separation between domain-dependent knowledge-base and generic inference engine (for example, in frame-based knowledge representations such as SRL there is no generic inference engine).

A list of differences between traditional programs and KBES is provided in [Adeli85]:

1. Expert systems are knowledge-intensive programs,

2. In expert systems, expert knowledge is usually divided into many separate rules,

3. The rules forming a knowledge base or expert knowledge is separated from the methods for applying the knowledge to the current problem. These methods are referred to as inference mechanism, reasoning mechanism, or rule interpreter,

4. Expert systems are highly interactive,

5. Expert systems have user-friendly/intelligent user interface, and

6. Expert systems to some extent mimic the decision making and reasoning process of human experts. They can provide advice, answer questions, and justify their conclusions.

Point 1. is frequently cited, but can be misleading. A better characterization is given by [Newell85]: an algorithmic program uses a small amount of knowledge (e.g., the "knowledge" of matrix multiplication) repeatedly over many cycles, whereas KBES typically has to search a large amount of knowledge at each cycle, and a particular piece of knowledge may apply only once. Point 2. is a consequence of point 1. Points 4. and 5. are not restricted to KBES; any program can be made highly interactive (at least, in its input/output) and very user friendly. The first part of point 6. is not very relevant; very few expert system developers have explicitly claimed this feature.

Extended Definition of KBES

By elimination from Adeli's list, two key features clearly separate a KBES from an algorithmic program:

1. Separation of knowledge-base and control. There are some facilities for manipulating the knowledge-base per se (displaying, searching, modifying) separate from the control (inference engine) which "executes" the knowledge-base. (This is Adeli's point 3., but it is extended here to include problem-solving formalisms other than rule-based, pattern-directed production systems.)

2. Transparency of dialog. There is some form of an explanation facility to convey to the user the inference process actually used. Conventional "Help" facilities do not qualify, as these are again separate from the actual execution. (This is Adeli's point 6.)

Two additional distinguishing features have to be added to the above two.

3. Transparency of knowledge representation. The domain-dependent knowledge incorporated in the program code is readable and understandable to some degree. Full natural language translation, as in EMYCIN and its derivatives, is not necessary; on the other hand, comments do not qualify, as they are not incorporated in the code.

4. Incremental growth capability. The KBES can be used with a subset of its ultimate knowledge base, and its knowledge base incrementally extended over a period of use without major (or any) restructuring.

Development Tools and Environments

A wide variety of expert system development tools and environments are available, and new development frameworks are announced every few months, especially for PC-based applications. For the purpose of this paper, two key aspects of these frameworks are of significance.

First, there is a large variety of formalisms or languages for defining the knowledge base. Even in the most popular formalism, that of IF-THEN rule-based representation of the knowledge-base, there is a great variety of knowledge representation, ranging from natural language-like statements to essentially procedural statements. As another exercise in historical analogy, the status of expert system knowledge representation is akin to the status of procedural knowledge representation before the emergence of FORTRAN and other high-level procedural languages. Readers who remember the late 1950's and early 1960's will recall the bewildering array of interpreters and translators developed at that time to facilitate program development. While it is too early to expect the immediate emergence of a knowledge representation language comparable to FORTRAN in its generality, user acceptance and portability, it is clear that such a language (or small set of languages) will become available in a few years.

Second, many of the expert-system frameworks, particularly the LISP-based ones that have their origin in AI research, incorporate exploratory programming environments which have little to do with expert systems per se. These environments provide facilities for incremental compiling, screen and file managements, tracing, debugging, etc. which are highly conducive to and supportive of developing programs in application areas where program specifications are not clear at the beginning. Some early expert system developers chose these environments because of the extensive support facilities provided, and not because of any "built-in" knowledge representation and processing capabilities. Many of the newer expert system development frameworks continue to provide this rich support environment for incremental construction and modification of expert systems.

The Standard Development Process

The current philosophy of expert system development, as exemplified by texts such as [Hayes-Roth83], is predicated on the assumption that an expert system is generated through the cooperation of two groups of individuals: one or more domain experts with experience and knowledge about the application domain, and one or more knowledge engineers, who are knowledgeable about several expert system frameworks, so that they can: a) choose an appropriate representation and inference strategy; b) guide the domain experts in developing the relevant knowledge base; and c) implement the experts' knowledge base in the framework selected.

This philosophy is, again, reminiscent of the state of computing before FORTRAN, where the domain expert had to rely on an intermediary, then called a programmer or coder, who could translate the domain expert's procedural knowledge into the then-available assemblers, interpreters, subroutines, etc.

Critique of Present Development Process

Even at this early stage in the history of expert systems, it is clear that the standard development process sketched above is not going to be the predominant mode of expert system development. First, there are just not enough conventional knowledge engineers coming out of Computer Science. Second, explaining specialized domain knowledge to someone with no background in the domain is probably harder than "explaining" it to a program, i.e. an expert system development framework. Third, the authors of such frameworks are beginning to provide user interface and knowledge acquisition facilities so as to eliminate the need for a human intermediary.

It is thus eminently clear that the next generation of expert systems will be written by today's application programmers, educated in their respective application domains, and capable of incorporating substantial segments of domain knowledge into their expert systems based on their own personal expertise. The need for additional expertise will be confined to higher-level issues beyond the direct expertise of the application programmer.

Expected Future

Today's expert systems do not exhibit any of the features of AI programs that can modify their behavior, learn by generating new rules, generalize from examples to higher abstractions, etc. In fact, most of today's expert systems, particularly rule based ones, deal only with shallow knowledge, i.e. empirical associations. Even when deep, i.e. causal, knowledge is used, it is treated the same say as empirical knowledge. The future use of KBES in civil engineering can be expected to expand in several directions.

First, the exploratory programming environment will become increasingly available in situations other than KBES development, as, for example, in TURBO Pascal, which will make program development easier and more congenial for everyone.

Second, KBES environments will be more closely coupled with algorithmic programs which can supply the deep, causal knowledge. Thus, KBES will be increasingly used not as standalone programs, but as intelligent pre- and post- processors for existing programs, such as finite element analyzers or CPM schedulers.

Third, KBES frameworks will provide increasing user interface, explanation, knowledge acquisition and imprecise reasoning facilities. As described earlier, the major users of these frameworks will be engineering application programmers, and not knowledge engineers in today's sense.

Finally, on a slightly longer time scale, further research results from AI will transfer into practical use, including some techniques for handling conflicting expertise and certain forms of automated learning.

In summary, the excitement created by the emergence of KBES may be largely unfounded, as judged by the first generation of KBES developed to date. Nevertheless, KBES offer a way to deal with the ill-structured aspects of civil engineering. The shortage of good, practical civil engineering KBES is due not so much to the limitations of present KBES frameworks as to the difficulty of compiling, organizing and formalizing the vast body of heuristic expertise which characterizes the profession.

References

[Adeli85]   Adeli, H. "Knowledge-Based Expert Systems in Structural Engineering", Proceedings of the Second International Conference on Civil and Structural Engineering Computing, London, United Kingdom, December 3-5, 1985.

[Hayes-Roth83]   Hayes-Roth, F., D. Waterman and D. Lenat, *Building Expert Systems*, Addison-Wesley Publishers, 1983.

[Gaschnig81]   Gaschnig, J., R. Reboh and J. Reiter, "Development of a Knowledge-Based System for Water Resources Problems", SRI Project 1619, SRI International, August, 1981.

[Newell85]   Newell, A., Private communication, 1985.

Problem Solving Using Expert System Techniques

Mary Lou Maher, A.M.ASCE[1]

1. Introduction

Computers have traditionally been used to solve engineering problems that are formalized and analytical in nature. Using conventional programming techniques, a list of sequentially executable statements must be formulated before the computer can solve the problem. This requirement for explicit formalization of the problem into detailed, sequential statements has restricted the use of the computer to problems that have solutions that are well understood. The desire to use the computer to aid in the solution of engineering problems that are less formalized or understood has led to the recent interest in expert system techniques. Problems are solved in expert systems by a set of rules and/or procedures whose execution order depends on the problem solving strategy utilized.

There are many alternative problem solving strategies that can be implemented using expert system tools and techniques. However, there are basically two approaches to problem solving currently used in expert systems: the derivation approach and the formation approach. The derivation approach involves deriving a solution that is most appropriate for the problem at hand from a list of predefined solutions stored in the knowledge base of the expert system. The formation approach involves forming a solution from the eligible solution components stored in the knowledge base. Depending on the complexity of the problem being solved, an expert system may use one or both of the approaches described above.

This paper begins with a description of a subset of the problem solving strategies used in expert systems and how these strategies support the derivation or formation approach. This is followed by the description of a structural design problem and its implementation using first a formation approach and then a derivation approach. Finally, some comments and conclusions are provided.

2. Problem Solving Strategies

Problem solving involves the search for a solution. The search begins at an initial state of known facts and conditions and ends at a goal state. The solution path consists of all states that lead from the initial sate to the goal state. In a formation approach to problem solving the known facts and conditions are combined to form a goal state. In a derivation approach, the known facts and conditions are used to derive the most appropriate goal state.

Domain independent problem solving strategies are commonly referred to as weak methods and may lead to combinatorial explosions due to a potential lack of focus. Expert systems can be considered strong problem solvers since they employ domain knowledge in the solution strategy. In this section a number of problem solving strategies currently used in expert systems are briefly presented and discussed in light of their potential for implementing a

---

[1] Assistant Professor of Civil Engineering, Carnegie-Mellon University Pittsburgh, PA 15213

formation or derivation approach. More detailed descriptions of a number of problem solving strategies can be found in [Nillson 80, Rich 83, Stefik 77], and a review of problem solving strategies appropriate for engineering design in [Maher 84a].

The following strategies are appropriate for the implementation of a derivation approach: forward chaining, backward chaining, and mixed initiative. These strategies require that the goal states represent the potential solutions and the initial state represent the input data. The use of these strategies require the development of an inference network representing the connections between initial states and goal states, as illustrated in Figure 1. The advantage to using one of these strategies is that they are currently implemented in a variety of expert system tools so that the development process involves defining, testing, and revising an inference network.

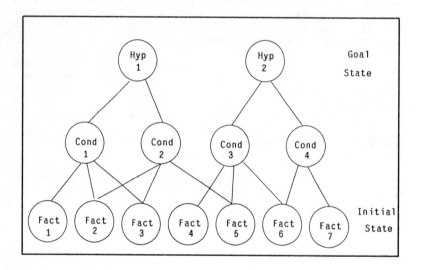

Figure 1: Inference Network For A Derivation Problem

Forward Chaining. A system uses a forward chaining (bottom-up, data-driven, antecedent-driven are all equivalent to forward chaining) strategy if it works from an initial state of known facts to a goal state. The input to such a system includes the value of all facts that the system has in its knowledge base. The main drawback of this strategy is that it is extremely wasteful to require as input data all the possible facts for all conditions; in many circumstances all possible facts are not known or relevant. This strategy is useful in situations where there are a large number of hypotheses and few input data.

Backward Chaining. A system uses a backward chaining (also referred to as top-down, goal-driven, hypothesis-driven and consequent-driven) strategy if it tries to support a goal state or hypothesis by checking known facts. If the known facts do not support the hypothesis then the preconditions that are needed for the hypothesis are set up as subgoals. This process continues until the original hypothesis is either supported or not supported by known facts. The system may then pursue the validity of another hypothesis in the knowledge base. The order in which the hypotheses are pursued is predefined.

Mixed Initiative. A system uses a mixed inititative strategy when it combines the forward chaining and backward chaining strategies. The system starts with the initial state of known facts to assign a probability to each of the potential goal states. The system then tries to support the goal state with the highest probability by setting up subgoals and requesting additional information from the user if necessary. In the mixed initiative strategy the order in which the hypotheses are checked depends on the problem at hand. The advantage to this strategy is that the user only supplies the data relevant to the problem at hand and not all possible data values.

The problem solving strategies appropriate for implementing a formation approach are: problem reduction, plan-generate-test, and agenda control. These strategies may be supplemented with the concepts of hierarchical planning and least commitment, backtracking, and constraint handling techniques. The development of an expert system using one of these strategies requires the definition of the components of the solution and a description of how the components can be combined. An illustration of the unconnected graph of components is shown in Figure 2. The solution is not completely defined by a goal state, but requires that the solution path be known also. The disadvantage to using one of these strategies is the lack of a standard implementation or expert system tool that employs a strategy appropriate for the formation approach. These strategies are typically implemented using a lower level language such as Lisp.

Problem Reduction. Problem reduction involves factoring problems into smaller subproblems. The problem is represented as an AND-OR graph. An AND node consists of arcs pointing to a number of successor nodes, all of which must be solved for the AND node to be true (or solved). For an OR node, it is sufficient for one of the successor nodes to be solved; an OR node indicates that a number of alternate solutions exist for the problem.

Plan-Generate-Test. The generate-and-test strategy in its pure form generates all possible solutions from components in the knowledge base and tests each solution until it finds a solution that satisfies the goal condition. The plan-generate-test sequence restricts the number of possible solutions generated by an early pruning of inconsistent solutions. The pruning is achieved by the planning stage, where the data is interpreted and constraints are evaluated; these constraints eliminate solutions that are inconsistent.

Agenda Control. The agenda control strategy involves assigning a priority rating to each task in the agenda. The task with the highest priority is performed first. A list of justifications and a method of determining the priority rating is associated with each task. This type of control can be used for complex tasks that require focusing attention on certain parts of the problem. Agendas can also be used in systems that require several independent sources of expertise to communicate with each other.

Hierarchical Planning & Least Commitment. The concept of hierarchical planning involves the development of a plan at successive levels of abstraction. For example, in the design of a complex system the design space is divided into a set of levels, where the higher levels are abstractions of the details at lower levels; the problem is hierarchically decomposed into

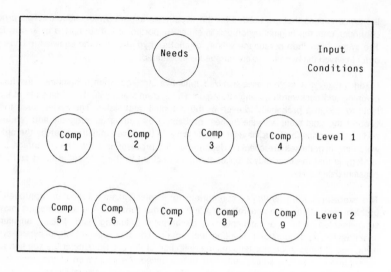

Figure 2: Unconnected Graph For A Formation Problem

loosely coupled subsystems. The least commitment principal involves deferring the assignment of values to variables until more information about the problem space is available. Least commitment is appropriate when a number of solutions exist for each subsystem and the solution of one subsystem depends on decisions made in the solution of another subsystem.

Backtracking. In backtracking the problem solver backs up to a previous level in the solution process if no solution is found along the current path. Backtracking is necessary when a number of alternate approaches to a problem solution exist. Backtracking is provided in some languages, such as PROLOG [Clocksin 81], but must be programmed in most languages. Backtracking, in its pure form, poses a number of difficulties. To provide an efficient way of backtracking from wrong guesses, Stallman and Sussman [Stallman 77] developed the concept of dependency-directed backtracking (DDB). In DDB, a record of all deduced facts, their antecedent facts along with their justifications are maintained; this information is known as dependency records. The records are used to drive the backtracking process when a contradiction occurs. This concept requires a lot of bookkeeping, but the additional bookkeeeping may be used for explanation purposes as well.

Constraint Handling. The interaction between subsystems can be handled by constraint satisfaction techniques. Constraint satisfaction techniques involve the determination of problem states that satisfy a given set of constraints. Stefik [Stefik 80] proposes three operations on constraints:

   1. *Constraint formulation* is the operation of adding new constraints representing

restrictions on variable bindings.   Typically, the constraints contain increasing detail as the solution progresses.

2. *Constraint propagation* is the operation of combining old constraints to form new constraints. This operation handles interactions between subproblems through the reformulation of constraints from different subproblems.

3. *Constraint satisfaction* is the operation of finding values for variables so that the constraints on these variables are satisfied.

3. Example: Configuring Alternative Lateral Load Resisting Systems

The example selected to illustrate the flexibility of the expert system approach to problem solving is a structural design problem. The ideas used to develop the problem where taken from two expert systems developed at Carnegie-Mellon University: HI-RISE [Maher 84b] and LOW-RISE [Camacho 85]. These expert systems represent an effort to use expert system techniques to study the preliminary structural design process. HI-RISE is an expert system that configures and evaluates alternative structural systems for high rise buildings with limited capabilities for spatial reasoning. LOW-RISE configures and evaluates alternative structural systems for low rise industrial buildings giving some consideration to spatial reasoning. HI-RISE uses a formation approach; the alternative structural systems are formulated by combining and proportioning structural subsystems and components. LOW-RISE uses a derivation approach; the structural components and subsystems are combined and stored in the knowledge base as alternative configurations. The example in this paper is a subset of the design performed by HI-RISE; the definition of feasible alternative lateral load resisting systems. The design will be implemented first using the formation approach that HI-RISE uses and then using the derivation approach that LOW-RISE uses.

3.1 Problem Statement

The problem is to define alternative feasible structural configurations for the lateral load resisting system for a given building. The input to the problem is a three dimensional grid representing the potential locations of structural subsystems and components, as shown in Figure 3. The output is a list of feasible structural alternatives.

The problem statement is further refined by defining the structural subsystems and components that are to be considered. The components are decomposed into hierarchical levels of alternatives. The three levels used for this example are 3D subsystems, 2D subsystems, and lateral system material. The alternatives in each level are shown in Figure 4.

The 3D subsystems are: tube, a three dimensional moment resisting structure placed on the perimeter of the building; core, a three dimensional moment resisting structure placed around an internal core; and orthogonal 2D, two dimensional moment resisting structures placed in orthogonal directions. The 2D subsystems are braced frame, rigid frame, and shear wall. The lateral system materials considered are steel and reinforced concrete.

Heuristics are used to determine whether the combination of alternative subsystems under consideration represents a feasible system for the given problem. Some heuristics are based on material subsystem compatibility. For example, the selections of steel as the material and shear wall as the 2D subsystem is not a feasible combination. Other heuristics are based on

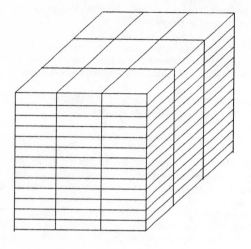

Figure 3: Input To Lateral System Configuration Problem

LEVEL:	RANGE:
3D Subsystem	tube, core, orthogonal 2D
2D Subsystem	braced frame, rigid frame, shear wall
Lateral Material	steel, reinforced concrete

Figure 4: Subsystem Levels For Lateral System Configuration Problem

experience derived from previous building designs. For example, the selection of tube as the 3D subsystem for a building less than 40 stories is not a structurally viable solution.

3.2 Formation Approach

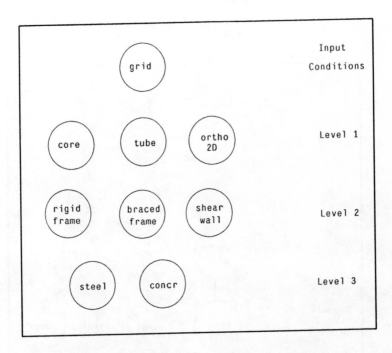

Figure 5: Unconnected Graph For Lateral System Design

The formation approach to the design problem involves the development of an unconnected graph to represent the hierarchical decomposition of the component alternatives, as shown in Figure 5. A generate and test strategy is used to define the feasible alternatives. The components are combined using a depth-first processing of the hierarchy, checking constraints at each level in the hierarchy.

The constraints associated with each level represent design heuristics. For example, some constraints associated with the 3D subsystem level are given below.

```
IF number of stories < 40, AND
   3D subsystem = tube
THEN eliminate alternative.

IF number of stories < 40, AND
   3D subsystem = core
```

```
THEN eliminate alternative.

IF number of stories > 40, AND
   3D subsystem = orthogonal 2D
THEN eliminate alternative.
```

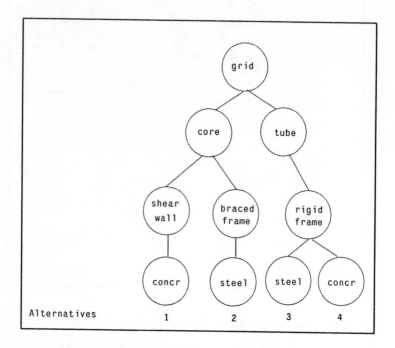

Figure 6: Tree Of Alternative Configurations

The design process proceeds by selecting one alternative from each level of the hierarchy and checking the elimination constraints. If the alternative is eliminated, another item is selected from the same level. If the alternative is not eliminated, an item is selected from the next level. The result is a tree of alternative configurations, representing the solution path, as shown in Figure 6. The figure illustrates the following alternatives: alternative 1 is a combination of a core, a shear wall, and reinforced concrete; alternative 2 is a combination of a core, a braced frame, and steel; alternative 3 is a combination of a tube, a rigid frame, and steel; and alternative 4 is a combination of a tube, a rigid frame, and reinforced concrete.

3.3 Derivation Approach

The derivation approach to the design problem involves the development of an inference network to represent the connections between feasible solutions and input data, as shown in Figure 7. The feasible solutions represent predefined alternative configurations. For example, alternative 1 represents a reinforced concrete shear wall around the core, alternative 2 represents a steel braced frame around the core, alternative 3 represents a steel rigid frame tube, and alternative 4 represents a reinforced concrete rigid frame tube.

The heuristics that were represented as elimination constraints in the formation approach are reresented as decisions in the inference network in the derivation approach. The design can be implemented using a forward chaining, backward chaining, or mixed initiative strategy. Using the backward chaining strategy, the first alternative is pursued by traversing the network to check if the input data can support the alternative. The other alternatives are pursued in turn. The solution is represented by the alternatives that can be supported by the input data.

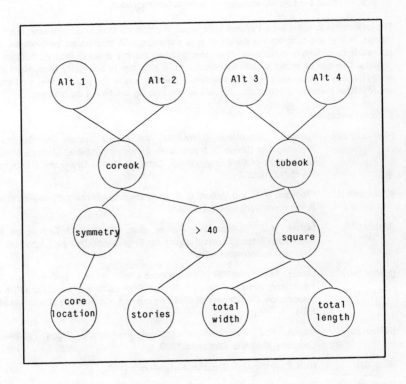

Figure 7: Inference Network For Lateral System Design

4. Summary and Conclusion

This paper presents a review of the most common problem solving strategies used in the development of expert systems in light of implementing either a derivation or formation approach to problem solving. The difference between a derivation and formation approach is illustrated by the implementation of a design problem using both approaches. It is important to note that the flexibility in using either a derivation or formation approach for a design problem is possible only when the design problem is simple. A complex design problem, involving design of multiple, interacting subsystems, cannot be easily implemented using a pure derivation approach.

The author's experience in developing expert system for structural design has been that the initial problem solving approach appropriate for design problems is a formation approach, as this approach is similar to the approach that humans use. Only after the definition of the relevant components and constraints on their combination using a formation approach is it possible to recast the solution strategy into a derivation approach.

The advantage to using expert system techniques in solving ill-structured problems, such as design, is that the problem solving strategy is transparent. A transparent problem solving approach is a definite advantage when the problem solving process may be subject to change. The use of expert system techniques allow changes to be readily recognized and relatively easy to implement. And as is shown in the example in this paper, using a different approach for the same problem can give new insight into the solution of the problem.

5. References

[Camacho 85]   Camacho, Godofredo, *LOW-RISE: An Expert System For Structural Planning And Design Of Industrial Buildings*, unpublished Master's Thesis, Department of Civil Engineering, Carnegie-Mellon University, Pittsburgh, PA 15213, 1985.

[Clocksin 81]   Clocksin, W. F. and Mellish, C. S., *Programming in Prolog*, Springer-Verlag, Berlin Heidelberg New York, 1981.

[Maher 84a]   Maher, M.L. , Sriram, D. , Fenves, S.J., "Tools and Techniques for Knowledge Based Expert Systems for Engineering Design," *Advances in Engineering Software*, 1984.

[Maher 84b]   Maher, M.L., *HI-RISE: A Knowledge-Based Expert System For The Preliminary Structural Design Of High Rise Buildings*, unpublished Ph.D. Dissertation, Department of Civil Engineering, Carnegie-Mellon University, 1984.

[Nillson 80]   Nillson, N. J., *Principles of Artificial Intelligence*, Tioga Publishing Company, Palo Alto, California, 1980.

[Rich 83]   Rich, E., *Artificial Intelligence*, McGraw Hill, 1983.

[Stallman 77]   Stallman, R., and Sussman, G. J., "Forward Reasoning and Dependency-directed Backtracking in a System for Computer-Aided Circuit Analysis," *Artificial Intelligence*, Vol. 9, pp. 135-196, 1977.

[Stefik 77]   Stefik, M. and Martin, N., *A Review of Knowledge Based Problem Solving as a Basis for a Genetics Experiment Designing System*, Technical Report STAN-CS-77-596, Computer Science Department, Stanford University, March 1977.

[Stefik 80]   Stefik, M., *Planning With Constraints*, Technical Report STAN-CS-80-784, Computer Science Department, Stanford University, January 1980.

Expert System Tools for Civil Engineering Applications

Phillip J. Ludvigsen, William J. Grenney[1], Del Dyreson[2], and Joseph M. Ferrara[3]

INTRODUCTION:

In the past, expert system development was a monumental undertaking reserved for major universities and corporate giants. Fortunately advances in microcomputers coupled with a more pragmatic understanding of how expert system technology can be applied have initiated a new era in user developed expert systems. Development time, that took multi-man years, now takes months if not weeks. Programming, which required highly technical computer skills, can now be accomplished by novice programmers with the aid of software tools. Accessibility to AI (artificial intelligence) expertise, that was only available on the university campus, is now available (if only indirectly) through customer service, support, and training. All these changes manifest in one more factor - the cost. Projects that once required major public and/or private funding can now be undertaken by small companies and even individuals.

Expert system software tools have played a major role in expediting program development, however, they do not offer a panacea to all problems which require expertise. It is important to know that some problems should not and possibly can not be solved by current expert system technology. For problems that can be solved with current technology, consideration must be given to the design of the tool and how it relates to your particular problem. Each tool, as with any software, has advantages vs. limitations which must be evaluated before project development begins. The final decision on which tool is "best" (most appropriate) is dictated by various factors, such as flexibility, user support, documentation, and of course cost.

[1] Graduate Research Assistant, Professor, Respectively, Dept. of Civil and Environmental Engineering, Utah State Univ., UMC 41, Logan, UT 84322

[2] Associate Professor, Computer Science Department, Utah State Univ., UMC 42, Logan, UT 84322

[3] Associate Research Professor, Artificial Intelligence Research & Development Unit, Dept. of Special Education, Utah State Univ., UMC 68, Logan, UT 84322

The purpose of this paper is not to endorse any one particular software tool to build civil engineering expert system applications, but rather to emphasize their particular advantages as well as limitations from a civil engineering standpoint. The information presented is vintage 1985 and one should be aware of current changes in program features and price. Caveats aside, the body of this paper is based on hands-on experience and should prove useful.

Brief History:

Expert systems tools, sometimes called authoring tools, or more recently shells, have a relatively short history. Basic research of developing a tool to aid in building an expert system is approximately ten to fifteen years old. Much of the early work was done in the field of medical diagnosis. From this work, a tool named MYCIN emerged. It used "if-then" production rules, certainty factors and backward-chaining inference, thus setting the standard which many current tools follow today (1).

Teknowledge (Palo Alto, CA), probably the largest and oldest "expert systems company" has a corporate history of approximately five years. Experience in developing expert system tools and applying them is somewhat limited, especially in the field of civil engineering. Fenves, Maher, and Sriram in their paper "Expert Systems: C.E. Potential" highlight future uses of expert systems but the lack of current applications is apparent (2).

It is interesting to note the evolution of these tools and how recent advancements affect solving problems within the civil engineering domain. But first, lets take a look at what distinguishes civil engineering problems from other kinds of complex problems.

Problem Domain:

Civil engineering exhibits an extremely wide, as well as deep problem domain. The sheer diversities of disciplines involved and complexities encountered are self-evident. Because of this, civil engineering expert systems and thus the tools to build them must be extremely flexible. The ideal tool for building civil engineering systems would allow for the following:

* Complex mathematical manipulations within the tool
    (Including scientific functions plus canned
        algorithms, ie. statistics, integration, etc.).
* Various forms of knowledge representation
    (Not just "if-then" production rules).
* Various inference strategies
    (Not just "backward-chaining").
* Simple calls to other programs or expert systems
    written in ANY programming language.
* Natural language interfaces.
* Unlimited degree of expert system explanation.

* Extensive development environments
    (eg. intelligent editors, debuggers, graphics,
    and help facilities).

Unfortunately, no tool available today allows for all of
the above. Many optimistic sales and customer service
people will tout "we can't do that directly, but we can
show you ways to work around it" or "our next version is
slated to have that improvement". Upon hearing such
statements, beware! Given the time and money one can
"work around" or wait for anything, but the time or money
might not always be available.
    A good example of what appears to be a universal
limitation of current tools is the inability to handle
complex mathematical manipulations directly within the
tool. The standard solution is to call (sometimes
referred to as "hook") a module written in some common
programming language to return the desired calculated
data. However, you might have to call this module or
others many times within an iterative solution process.
This can slow execution down tremendously. In some cases,
you can use knowledge engineering (programming) "tricks"
for greater efficiency. The drawback to this approach is
that your solution logic is dictated more by the tool's
limitations than by the problem.
    Even though current tools fall short of the ideal,
the future looks bright. Expert systems and the tools to
build them are heavily dependant on hardware speed and
memory. This is why we see special machines designed just
for AI work. Fortunately, advancements in hardware design
are bringing tremendous computing power and thus making
meaningful expert system tools available to desk top
computers.

Tools:
    Seven tools are investigated (EXPERT-EASE, INSIGHT,
M.1, RuleMaster, EXPERT, ROSIE, and S.1). They represent
a diversity of complexity, flexibility and cost.
EXPERT-EASE, INSIGHT, and M.1 are suitable for micro
computers while RuleMaster is a medium size tool suitable
for super mini's. EXPERT, S.1, and ROSIE can be considered
large, main frame software.
    The following criteria are used to investigate
important features of each tool.

* Approximate cost.
* Ability to handle complex mathematics.
* Ability to interface with other software.
* Explanation facilities.
* Overall friendliness.
* Documentation.
* User Support.

As means of a brief summary, Table 1 compares each
evaluated tool according to the above criteria.   The

EVALUATED EXPERT SYSTEMS TOOLS

ATTRIBUTE	SMALL			MEDIUM		LARGE	
	EXPERT-EASE	INSIGHT II	M.1	RULEMASTER	EXPERT	ROSIE	S.1
Proprietary Interest	Human Edge Software (Palo Alto, CA)	Level Five Research Inc. (Melbourne Beach, FL)	Teknowledge (Palo Alto, CA)	Radian Corp. (Austin, TX)	Rutgers Univ. (New Brunswick, NJ)	Rand Corp. (Santa Monica CA)	Teknowledge (Palo Alto, CA)
Phone	800-624-5227	305-729-9046	415-327-6640	512-454-4797		213-393-0411	415-327-6640
Approx. Cost	$695 (IBM-PC)	$495 (IBM-PC)	$5000 (IBM-PC)	$10,000 (IBM-PC) $25,000 (VAX)	No Charge	$200 (VAX)	$50,000 (VAX)
Ability to handle Complex Math	NO	YES (directly)	YES (indirectly)	YES (indirectly)	NO	NO	YES (indirectly)
Ability to interface with other software	NO	YES (directly)	YES (indirectly)	YES (directly)	NO	NO	YES (indirectly)
Explanation	NONE	Extensive	Limited	Extensive	Limited	Very Limited	Extensive
Overall Friendliness	Buddies	?	A Friend	A Friend	Fair-Weather Friend	Fair-Weather Friend	Real Buddies
Documentation	Outstanding	?	Sufficient	Meager	Meager	Sufficient	Sufficient
User Support	Limited	Available	Available	Available	None	None	Available

Table 1. Expert system tools comparison.

following offers a bit more detail.

Expert-Ease:

Expert-Ease (Expert Software International Ltd) is probably the smallest of the tools evaluated. It is designed to aid the use in quick development of small prototypes. This Pascal based tool features an automatic induction routine. One sets up a decision table and Expert-Ease translates it to Pascal code which can only be executed from within Expert-Ease. The user really has very little control over the inference strategy (dedicated forward-chaining). If Expert-Ease sees fit to ask a certain question first, the programmer (knowledge engineer) can not get at the Pascal code to over-ride the tool's decision.

This tool does not handle mathematical functions nor will it run on several IBM compatible machines. Expert-Ease does not make allowances for interfacing to other software, plus it does not have any explanation facilities. Due to these limitations, Expert-Ease can not be considered for large complex problems which involve mathematics, this rules out most engineering problems. However, Expert-Ease could be used to develop skeletal logic structures involved in solving larger problems. It has outstanding documentation and it is easy to use. But for large engineering expert systems this tool will not handle the load.

Other descriptions and evaluations of Expert-Ease are available (3, 4, 5).

Insight II:

Insight II (Level Five Research) is the newest of all the tools evaluated. Unfortunately, we were not able to receive an evaluation copy in time for this paper. So, attributes such as overall, friendliness and quality of documentation are not evaluated here. Without hands on experience with Insight II, little can be said about its limitations and shortcommings, however, Insight II appers to have some powerful facilities at an attractive price ($495).

Insight II is a Pascal based program which boasts a menu - driven development environment, links to other programs, confidence factors, tiered explanation, the ability to produce "run-only" end user versions, the ability to address 2000 rules, the speed of a compiler based system, and believe it nor not, complex mathematical functions which are intrinsic to the system. Insight II is the only tool evaluated which directly incorporates transcendental math functions (ie. cos, sin, tan, etc.). For such a low price, this tool might be just the ticket, but just how well everything fits together remains to be seen.

M.l:

M.1 (Teknowledge Inc.) is considered by many to be

the Cadillac of PC-based expert system tools. The reasons
for this analogy is 1) its price ($5000 - down from last
year's price of $10,000); 2) it has many features that
were once found on only mainframe system tools; and 3)
Teknowledge offers first class user support (training and
consulting). Teknowledge has gone to great lengths to
produce a professional piece of software, but just as a
Cadillac can be inappropriate for certain jobs, so can
M.1.

M.1 (version 1.3) is a Prolog based tool which
interprets english-like production rules. The form of
inference used is backward chaining. Forward chaining can
be simulated but the result is somewhat awkward. Because
M.1 acts as an interpreter, rules are acted on much slower
than a compiler based system. This can be a limiting
factor, especially if you have to constantly "hook" out to
other software. M.1 can not handle complex mathematics or
interface with other software directly from within itself.
One must customize the provided interfaces, write
interfacing software in assembly language or "C" (Oh
Boy!), and finally link the whole thing together with M.1
to get a new executable version. This is not a trivial
task, considering most civil engineers are not fluent in
assembly language or "C".

The explanation facilities are somewhat limited.
When asked why?, M.1 will either give a programmer
supplied explanation of just the rule which caused the
query, or it will trace the rules (only by number) which
were used to reach that point in the run. Since most
expert systems reason within networks, it is impossible
for a knowledge engineer to write one single explanation
of a particular rule that will be in context. Some tools
use a "tiered" explanation in an attempt to establish
context, M.1 does not.

M.1 does not have an extensive development
environment, in fact it does not have its own editor to
make permanent changes or addition to the knowledge base.
One must leave M.1 in order to use your own text editor or
word processor. But M.1 does have nice tracing and
debugging facilities. It also has the ability to produce
"run-only" end user systems. This feature is particularly
attractive to those who wish to diseminate their work but
can not afford to buy numerious copies of the expert
system tool.

The documentation is sufficient but not impressive
when one considers the cost of the tool. About half of
the documentation contains example expert systems. These
examples are nice to have, but it would be nicer if each
M.1 command was defined along with numerous examples of
just how that command might be used. Included is a
helpful section on how to develop an expert system from
proposal to turn-key delivery. This "how to" section
offers some good program development advice, unfortunately
most of this rather large section has very little to do
with using M.1 directly.

There is some question whether or not M.1 can be used to build a significant expert system (>500 rules). The answer is maybe. M.1 can not address more than 200 rules at one time however, allowances are made to "shuffle" in and out groups of rules as needed. Here again, the interpreter nature of M.1 makes this process painfully slow.

M.1 is a powerful but expensive tool. Complex systems can be built but certain inconveniences (interfacing difficulties and slow execution) must be considered limiting. We are anxiously awaiting version 1.4 due out in early 1986.

Other descriptions and evaluations of M.1 are available (6, 7, 8).

RuleMaster:

RuleMaster (Radian corp.) is somewhat of an enigma. The tool is not just one program but rather three distinct entities: 1) Radial is a highly structured, pascal-like language; 2) Rulemaker is an induction routine which translates "examples" (logic tables) into Radial code (similar to Expert-Ease); and 3) the User interface, a sophisticated menu-driven collection of editors, tools, and various applications which help in the building of RuleMaster programs. What is puzzling about RuleMaster is its apparent lack of a separate control structure (inference engine). It is generally accepted that a separate control structure is one of the things that make an expert system - an expert system (9). If one works with RuleMaster, it becomes clear that the control structure is up to Rulemaker and/or the programmer (Knowledge engineer). This aspect of RuleMaster is truly a double-edge sword. On the positive side, the programmer is forced to structure the expert knowledge into modules that are easily updated by Rulemaker as well as the programmer. On the other side, trying to produce the effect of anything besides forward-chaining (ie. backward chaining) is practically impossible. Fortunately, many engineering problems are well structured and can be solved via forward-chaining inference.

RuleMaster handles complex mathematical functions by means of the usual "hook" to a separate program. What sets RuleMaster apart from the other tools evaluated is its ability to run under Unix or Unix-like operating systems. Since Unix can handle multitasking I/O, information can be easily shared between any number of programs written in any language. This is important if one has a large number of engineering algorithms written in different languages.

The explanation facilities are good. A nice feature is tiered explanation. One can keep querying the computer to get various levels of detail to the question - why?. RuleMaster does a good job at translating example knowledge to understandable explanation in context. The fact that the explanation can be put into context is not a trivial feature. If the problem is very complex, then a simple explanation of the rule being queried is easily misunderstood. The User interface makes working with this tool very enjoyable, but, the interface was just moving out of development in August 1985. At that time, there were some bugs and some of the applications were not available.

If RuleMaster has a soft-spot, it must be considered the lack of comprehensive documentation. For such a powerful tool, all we received is the training course notes which they call a "system user manual". User support is available through training and contract consulting.

The approximate cost of RuleMaster is $10,000 - $15,000 for IBM - PC and AT computers and $25,000 for supermini's such as a VAX. Educational discounts and trial period arrangements are available.

Important to note is that Radian is primarily a scientific - engineering company and its product RuleMaster is primarily designed for those domains. It is nice to know that if you do have problems, you can talk to a civil engineer who knows knowledge engineering rather than just a knowledge engineer who is not a civil engineer. All in all, RuleMaster is an excellent tool for building large systems that do not require various levels of abstraction.

Additional descriptive information is available from Radian Corporation (10).

EXPERT:

EXPERT was developed at Rutgers University for use in their biomedicine program so it is not surprising to find the tool's design directed at this domain. But, just as the field of medicine relies on expert diagnosis so do many problems in engineering. For example, diagnosing operational problems in a "sick" wastewater treatment system or aiding in structural design problems. Surprisingly, EXPERT is a Fortran based tool. It uses standard productions rules to represent procedural knowledge and in the tradition of MYCIN, EXPERT incorporates certainty factors within a backward-chaining inference.

Even though EXPERT is written in Fortran it only handles the standard mathematical operations of $+, -, /, *$, and $**$. It also does not hook out to other software (even Fortran) and unless you are a Fortran Guru with lots of extra time, do not expect to change this. The explanation facilities are somewhat limited. The user can ask why?, unfortunately the explanation only concerns the last question asked by the computer and not with the context of the question. EXPERT does have a "trace" facility which allows the user to follow the program's logic. For a large mainframe system, EXPERT's documentation is a meager 40 pages. This consists of an overview, a simple diagnostic example, and command definitions.

The main advantage of expert is its ability to run on larger computer systems (ours is running on a VAX 11-780). This allows for large number of rules to be incorporated and executed quickly. EXPERT is available for prototyping at little or no cost, however, there is no formal source of user support. EXPERT could be customized into a very powerful engineering tool, if one has access to a relatively large computer and is proficient in Fortran and fundamental expert system programming. If you can not afford customization, then plan on using EXPERT to solve diagnostic problems which do not involve complex algorithms.

Other descriptions and evaluations of EXPERT are available (11, 12).

ROSIE:

ROSIE (Rand Corporation) has been described as a general- purpose AI language as well as an expert system building tool (13). Since ROSIE is written in the INTERLISP programming language, it naturally picked up many of INTERLISP's features. As an expert system tool, ROSIE uses an English-oriented syntax in its knowledge base and input/output facilities. At first glance, it seems ROSIE has a built in natural language interface, however, it makes no attempt to grasp unrestricted English input. One must learn to "talk" to ROSIE in a very structured manner which resembles simple english sentences. It is still impressive to see the user interact with ROSIE by typing in small reports describing certain situations rather than answering one query at a time.

Because ROSIE is a large program which requires a large language (INTERLISP) to run, it commandeers significant memory and run-time. If one is paying for these services, the development costs can be restrictive. A mitigating factor is ROSIE (VAX-VMS version) can be obtained for approximately $200, but be prepared to spend several thousand dollars for INTERLISP.

ROSIE does not incorporate complex mathematical functions as part of the tool nor does it make allowances for interfacing with other software. The explanation facilities are very limited. One must use a "trace" or "scan" command to indirectly find out what is going on, rather than just simply asking why?. The development environment is also limited. There are no build in editors, menus, or graphical aids. The documentation consists of three volumes and is sufficient to get started on small to medium size systems. Little advice is given on building large systems via ROSIE. Finally, no formal support is available for ROSIE. Rand Corporation does consult on ROSIE but does not support it in a marketable way.

Additional descriptive information is available from Rand Corporation (14, 15, 16).

S.1:

Just as M.1 was considered the Cadillac of small system tools, S.1 (Teknowledge) is the Rolls Royce of the large system tools evaluated. The features of S.1 are too numerous to even list in this paper. Ironically, S.1's biggest drawback is its luxury. Just as most people can not afford to use a Rolls Royce as a pickup truck, most knowledge engineers can not afford to use S.1 to develop small systems. Unfortunately, most prototyping falls into this category. S.1 is a huge program that requires alot of computer memory and time to run. It had no trouble eating one of our time-shared VAX11-780's for lunch, in fact, S.1 should really have its own super mini dedicated just for itself (ie. Xerox 1108 workstation).

On one hand, S.1 has a sophisticated development environment consisting of an editor, debugging tools, and graphical aids. It allows for various forms of knowledge representation as well as inference strategies. Also, its explanation facilities are quite extensive. One can ask such question as how, what, and why. On the other hand, complex mathematical algorithms must be written outside of S.1 in the Lisp programming language (What fun!?). Also, interfacing to existing data-bases does not appear to be a simple task.

The documentation consists of five volumes, mostly training material or sample expert systems. Unfortunately there is no master index and just like M.1, no individual examples of each command are given. Here again, one must dissect an entire expert system to understand why and how certain commands were used in order to build ones own system.

Teknowledge is known for their outstanding user support, little of which comes for free, but still it's nice to know its there when needed. These people probably have more experience in building expert systems than any other company. Their products might cost more, but the strong user support might more than compensate for the initial investment.

One gets the feeling S.1 was designed to solve very grandiose but not very technical problems, something like automating the mail room at the pentagon. If price is no object and one is fluent in Lisp, S.1 allows enough flexibility to handle even engineering problems, but remember to ask yourself "do I need a pickup or a Rolls Royce". Needless to say, S.1 is not for everyone.

Case Study:

Utah State University, department of Civil and Environmental Engineering is pioneering the application of expert system technology to the areas of environmental systems modeling and hazardous waste management. One current study deals with the development of a demonstration expert sytem for assessing organic chemical mobility and degradation in order to consult on soil treatment options.

Since this system would be used for demonstration purposes, the portability of a PC-based program appeared attractive. For this same reason of demonstration, the ability to produce "run-only" versions of the expert system was considered an important factor. After some deliberation, M.1 was choosen for system development because the potential was there for building an expert system with the forementioned characteristics. However, one must be aware of the surrounding circumstances. First, since we are an education institution, all of the evaluated tools have been aquired at a greatly reduced cost or no charge at all. Second, Utah State University has available considerable in-house expertise in building expert systems as well as software engineering in general. Lastly, some of the newer PC-based tools, such as Insight II, were not available at the onset of this project. If carried out today, the decision of which tool to use for building this demonstration expert system would probably be different.

Food For Thought:

It is of utmost importance for any civil engineer who wishes to build an expert system to realize that one can learn to be a knowledge engineer rather rapidly, in fact many civil engineers are already knowledge engineers without even knowing it. Most engineers are quite good at extracting complex knowledge and based on scientific assumptions, produce simplified heuristics (Rules-of-thumb). Obviously, one can not be just a knowledge engineer and expect to become a civil engineer overnight. For this reason, civil engineers should seriously consider building their own systems with the help of flexible and user friendly tools before hiring those who are not familiar with civil engineering problems.

References:

1. Barr, A. and Feigenbaum E.A., "The Handbook of Artificial Intelligence, Vol 2," Heuristech Press, Stanford, CA, 1982, pp. 184-192.

2. Fenves, S.J., Maher, M.L., and Sriram, D., "Expert Systems: C.E. Potential," Civil Engineering/ASCE, Oct., 1984, pp. 44-47.

3. Derfler, F.J., "Expert-Ease Makes Its Own Rules," PC Magazine, April 16, 1985, pp. 119 - 124.

4. Pountain, D., "Computers As Consultants," BYTE, Oct., 1985, p. 367.

5. Webster R., "Expert Systems on Microcomputers," Computers and electronics, March, 1985, pp. 72-73.

6. Ibid., pp. 70-72.

7. Webster, R., "M.1 Makes a Direct Hit," PC Magazine, April 16, 1985, pp. 151-157.

8. Fersko-Weiss, H., "Expert Systems, Decision-Making Power," Personal Computing, Nov., 1985, pp. 99-105.

9. Duda, R.O, "Knowledge-Based Expert Systems Come of Age," BYTE, Vol. 6, No. 9, 1981, pp. 238-281.

10. Michie, D., et al., "RuleMaster, A Second-Generation Knowledge Engineering Facility," Radian Technical Report, MI-R-623, Dec., 1984.

11. Waterman, D., and Hayes-Roth, F., "An Investigation of Tools for Building Expert Systems," Prepared for the National Science Foundation, Report R-2818-NSF, 1982

12. Hayes-Roth, F., Waterman, D., and Lenat, D., "Building Expert Systems," Addison-Wesley Inc., Reading, MA, 1983, pp. 91-126.

13. Ibid, pp. 321-326.

14. Fain, J., et al., "Rational Motivation for ROSIE," A Rand Note, N-1648-ARPA, Nov., 1981.

15. Fain, J., et al., "The ROSIE Language Reference Manual," A Rand Note, N-1647-ARPA, Dec., 1981.

16. Fain, J., et al., "Programming in ROSIE: An Introduction by means of Examples," A Rand Note, N-1646-ARPA, Feb., 1982.

ATTRIBUTES AND CHARACTERISTICS OF EXPERT SYSTEMS

Celal N. Kostem*, M.ASCE

Abstract

Some of the characteristics of expert systems, with emphasis on the incorporation of these attributes in the development stage, are discussed. It is suggested that the scope of expert systems should be defined by the experts. Through the discussion of a number of case studies conclusions are drawn to be considered when developing the expert systems.

Introduction

This paper is not intended to present tools that can be directly used in the "design" of expert systems - referred to as ES hereinafter; but, to discuss some characteristics and attributes of ES. These characteristics are listed in most publications (e.g. Refs. 4 and 9); however, they are rarely examined in-depth in the presentation of finished products.

The paper cites a number of case studies, without identifying the parties involved, to draw conclusions on what should be done and what should not be done in the development and the use of ES. Most of the case studies given are taken from less than successful experience with ES. The thrust of the paper is not to criticize or condemn ES via these case studies; far from it. Just like the contributions of forensic engineering and pathology to civil engineering and medicine, respectively, the paper attempts to draw lessons from past "experiences."

---

* Professor of Civil Engineering, Fritz Engineering Laboratory, 13, Lehigh University, Bethlehem, PA 18015.

Expectations

During the last couple of years ES, and especially "artificial intelligence," AI in-short, as applied to various problems, have become the buzz-words for all professional, and general publications (e.g. Refs. 1, 6, 11, and 12). The writers of these feature articles and/or news are not always sufficiently well-versed in civil engineering and/or ES. These feature articles predict the solution of most problems via ES and AI techniques. Furthermore, ES or AI software developers who are not fully cognizant of the complexities of ES or AI or, worse, who are not sufficiently familiar with the specific civil engineering subdisciplines, have been prematurely announcing the pending release of packages yet to be completed, or developed. It is noted, perhaps with unwarranted pessimism, that some ES software developers have spent more time preparing press-releases, than actually working on the software system.

The high expectations placed on ES within the immediate future are noted above. Both the magnitude of the expectations and the timing can be best illustrated by a recent experience. A renowned engineer who used to manage a large number of engineers and computer systems wished to obtain a "routine" wordprocessor for his new practice. Recognition of the fact that the spelling-checker program can not differentiate between the proper use of "than" vs. "then" led him to conclude that as far as he was concerned the wordprocessor is not any better than a typewriter. Furthermore, he had expected that since the wordprocessors are so highly publicized and widely used, that these programs would be capable of catching grammatical errors, and correcting them. The write-ups that were glanced through also indicated to him the presence of the "language understanding" software (e.g. Refs. 8 and 10). As of the time of the writing of this paper, the professional correspondence of this engineer is still dependent upon his old typewriter.

To the uninitiated, the high-degree of publicity about the capabilities of ES and AI will inevitably lead to disappointment and alienation. If the civil engineering and ES professionals are to adopt realistic goals and timetables, than ES can be an integral part of civil engineering. Fenves, Maher and Sriram lucidly defined a number of potential areas in civil engineering for the application of ES (Ref. 3). The user community, after almost 20 years of exposure and experience, has an appreciation for what the finite element programs can and can not do. ES in civil engineering is relatively a more recent involvement. Until the user community develops a sufficient understanding of ES, it would be highly desirable for the developers not only to highlight the capabilities of any system, but also to specify its limitations. This does not detract from the value of the system, but it shows the professionalism of the developers. Most of the misunderstandings regarding the capabilities of ES and AI have been due to the incomplete definition of the packages by omission, rather than commission. It can hereby be concluded that:

    For any ES under development or soon to be released the
    developers must identify the inherent limitations of the

packages.

Expert System Models

In order to develop an ES it is imperative to have direct access to unbiased domain-specific "expertise." In the development of ES it is necessary for the expert and the knowledge engineer to interact and to be able to communicate (Ref. 4). Five examples are given to illustrate the various facets of ES, and how the developers can make mistakes.

In two separate cases two individuals highly knowledgeable on AI, and with very good understanding of structural engineering, but with no background on the finite element method, wished to develop expert systems for "universal," and "fully automated" finite element mesh-generators for plate bending problems. It was expected that the plate structures can have any shape, e.g. planar plates, folded-plates, etc., and can be stiffened by any structural component, e.g. concentric or eccentric beams, diaphragms, etc., and in addition, can be used for any "type" of problem, e.g. elastic or inelastic behavior, crack initiation and propagation, etc. The time estimates by the AI specialists for the envisioned projects were about one and three man-months. Needless to say, the projects never got off the ground. However, another project was successfully undertaken to generate finite element meshes for simple-span multi-beam highway bridges (Ref. 5). These generic structures correspond to planar rectangular plates with one-directional eccentrically placed stiffeners. The limited scope of the later project, as compared to the ambitious nature of the earlier, is the primary reason for its completion.

Another ES project, which was abandoned after extensive deliberations, was to develop software to find a scalar quantity "s," for given values of "I" and "J." The scalar quantity was uniquely defined by a matrix, i.e. s=S(I,J). This "table look-up" problem, because of the unique logic-path, does not require the deployment of ES and AI techniques, but a mere application of the "algorithmic programming."

> The scope of ES projects must be defined by the experts, rather than the knowledge engineers.

> The proposed ES projects with highly ambitious goals should be modularized, and the systems should be incrementally developed. In this development process, the flexibility and the generality of the systems should be maintained to permit the incorporation of future modules.

An ES package near completion drew criticism from the test-site users because of the extensive input required. The bulk of the required input could have been incorporated into the system by the developers. This was a tedious chore at best for the software developers. However, should this universal data base have been incorporated by the developers into the package, then the input required by the users would have been substantially reduced. Since "data" and "knowledge" are used interchangeably, expectation of the "pertinent data" to be inputted by the user means that the system is not knowledgeable enough, i.e. the system can be considered incomplete.

All ES must contain the required universally accepted knowledge-base embedded in the system. The lack of such a knowledge-base is the best indication of the incompleteness of the system.

An ES near completion created a high degree of controversy amongst the testers. This was due to the fact that the opinions of the experts on the subject area were clearly divided into two contradictory "approaches" for the type of problem that the ES was designed to handle. There, also, had not been sufficient experience on the success-rate of any of the advocated approaches. In order to get the ES development moving, one school of thought was embedded in the system. However, no "warning" was included in the ES to indicate the possibility that another approach could also be true.

Another ES under development created a controversy amongst the "experts" providing domain-dependent guidelines and information to be incorporated into the system. The experts were of the opinion that on the subject area knowledge is incomplete and sketchy in defining the solution to the problem. This system could have been designed in such a way that all accepted rules could have been embedded in the system, and the rules that are "speculative" could have been left out. As the new "rules" are developed and accepted, they could and should be incorporated into the system. Thus the system under development should have been labeled a "near-empty expert shell" or "semi-empty expert shell," as opposed to domain specific complete ES.

A complete ES should not be based on an incomplete, and/or an unaccepted knowledge-base.

Programming Language

A major manufacturer of climate-controlled "prefabricated buildings," e.g. large freezers, was designing and building one or two story high steel framing for these units. The size of these units varied from small, e.g. walk-in freezer, to essentially "unlimited" sized floor plans. All units were rectalinear in plan. The units were built essentially anywhere in the world, and they were subjected to all kinds of vertical and lateral loadings. In the tropics sometimes the governing roof loading was the wind induced suction, whereas in the northern US and Canada it was the snow accumulation. To expedite the bidding for the turn-key buildings, it was decided to have steel framing predesigned for any size building and for any loading combinations. It was further noted that the optimal steel cross sections can not be available on short notice in all locations. Thus a series of design alternatives had to be provided, starting with the optimal, all the way to totally uneconomical, as far as the steel use is concerned.

The above proprietary project, conducted by the author and two of his colleagues, resulted in a computer program to perform the design. The program provided all the steel sections, including the total tonnage in given structural shapes, in one short terminal session. The owner felt quite uneasy logging into a computer to input about half a dozen essential design data. The reluctance of the owner to do interactive computing necessitated the execution of the computer program for all possible combinations of geometries and loads. The total printout was in excess of 10 inches thick, and has been extensively used by the owner.

The above program contains all the prerequisites of an ES: The codification of expertise on a particular area, definition of the design parameters in simplistic terms, automated analysis and optimal design of the structural system, generation of alternate designs, etc. In view of the current ES or AI practice, it could be surmised that the above software is coded using a natural language. In reality the software was written in 1973-74 using FORTRAN-66, and absolutely no consideration was given to the maxims of AI. The program pre-dates the majority of the papers on ES or even AI. As was so succinctly noted by Fenves (Ref. 2), a number of programs written earlier could very well be referred to as ES, or at least pseudo-ES, codes. Conversely, some of the recent ES coded in declarative languages could very well have been written in FORTRAN or BASIC.

The use of languages like LISP, PROLOG, and OPS5 in the development of ES "simplifies" the coding. The major attribute of these languages, as compared to FORTRAN, is the simplicity of the addition, elimination, or substitution of new rules. This operation in high level languages like FORTRAN can be very simple, but it can also correspond to a major task. Insistence on the use of a declarative language, and labeling systems as being non-ES if they are written in a non-declarative language, is presumptuous.

> ES should be written in the most convenient language, with very strong preference given to declarative languages.

Declarative Languages vs. Empty Expert Shells

For many classes of problems it is possible to find "empty expert system shells" (Ref. 7). The price and the capabilities of these packages can vary from inexpensive to exorbitant, and from inappropriate to very powerful, respectively. For a large majority of prospective ES concepts and missions a knowledgeable "customer" should be able to find an empty expert system package. Incorporation of the rules to such a package is far simpler than to start with a natural language, and build an inference engine, etc. It is recognized that not all "problems" are amenable to the use of widely available empty expert systems. But it is observed that many other problems, that could have been handled via an empty expert system, are handled using languages like PROLOG or LISP. Probably the main factor for the use of natural language has been the no-cost availability of the natural language on a computer system, as opposed to $600-$30,000 investment required to obtain an empty expert system.

Feasibility to use an empty expert system must be investigated prior to programming in natural or high level languages.

Open vs. Closed Systems

Just as the name implies, expert systems have expertise built-in to them. Subconsciously the user community tends to have greater faith in an expert system, as compared to a general purpose algorithmic program written by a programmer of even very high repute. The possible easy access to the innards of an ES, by making changes in the "rules," can turn this finely tuned system not into a "garbage-in, garbage-out," but to "correct information-in, garbage-out." It is not unrealistic to expect that users of ES, with some degree of competence and confidence, will experiment with famous "What if..?" approach by changing some of the rules of the system, to identify the contribution of a given rule on the response of the ES. Thus, consideration is in order to define who should be permitted to make changes in the rules, and how.

Algorithmic programs are transmitted either in the form of source code or as a load module. In the latter case, the user can not make changes to the program. In the age of microcomputers limitation of the transactions to the load modules is sometimes labeled as a measure taken by the software developer to "insure" the clients or recipient's dependence on the developer for future interactions and transactions. However, to limit access to the source code, and especially to control the modifications to the source code, have benefits. The author's experience with those who have tried to modify source codes either to run in a new computer configuration, or much worse, to "improve and enhance" the code have been extremely disappointing. Those who do have a full understanding of the workings of the source code have always been cautious and employed defensive programming techniques. After every modification professional tests are carried out, and if the program does not yield the correct results, the changes are removed.

Less knowledgeable programmers tend to be more daring and they aggravate the errors with additional errors. The result of this action would be either the program (a) will give "slightly" incorrect results, or better (b) will give absurd results, or even better (c) will not compile and/or execute. The latter two cases should prevent blunders. The first case may have disastrous consequences. Under the circumstances the following categories are suggested for the expert systems.

Open Systems: An ES written in a declarative language, e.g. PROLOG, or in a high level language, e.g. FORTRAN, can be altered at the pleasure of the user to perform operations that the original version of the system can not perform. The availability of such a system for production runs by average uninhibited and uninformed users should be avoided at all costs.

Empty ES Shells: These are the ideal tools for expert system development. A desirable attribute of these systems would be to play back all added rules that have had direct effect on the decision making process. Furthermore, it would be most appropriate to have all the additions to the system dumped in frequent intervals. This would permit inspection of the rules by the experts for their appropriateness.

Completed ES Shells: An empty expert system shell in the hand of a true expert can turn into a powerful tool. As was indicated earlier it is also possible to have some rules incorporated into the system, and the others to be introduced at a later time. This possibility alone raises a "security" question. The rules could be added via the following approaches:

(a) The new rules can be fully incorporated by approval of the person(s) in-charge.
(b) No new rule can be permanently incorporated into the system. The user can introduce the new rules for given computer runs. The system will not incorporate these rules permanently, until they are "authorized" by the "system managers" (see above). The new rules employed should be highlighted on the printout or the file of the full dialog of the session.

The use of expert systems is encouraged, (if it is not, it should be), both for training and for production purposes. A careless user may very well introduce some "unacceptable" rules. The damage-control via identification of changes made to the basic officially accepted version of ES would be an excellent mechanism for ES-managers.

Any and all changes to ES packages should be carefully monitored. One or more individuals who are fully familiar with the ES should decide on any permanent or semi-permanent changes.

Calibration of Expert Systems

Initially some expert systems may not provide results "close enough" to those obtained from the field or laboratory tests. There exist several possibilities for such discrepancies: (a) the analytical model may contain errors, (b) the analytical model may not be appropriate to the types of problems on-hand, (c) experimental results may not be correct, and (d) due to various factors, too numerous to be listed herein, _exact_ duplication of tests by analytical models is "impractical." The first two reasons should and could be investigated by the ES developers. The last source of discrepancy is inevitable. Thus, the ES developers and users should learn to live with the "tolerable and acceptable differences." Unfortunately, engineers have far greater faith in experimental results, than those obtained by computer-based analytical models.

A developed system was successfully tested for all known "test cases." The system also had the option of considering various types of "symmetry" for the problems investigated. The symmetry option was incorporated to reduce the size of the problem. Just prior to release the symmetry options were tested by the project manager. The results were not the same as the full model. The primary worker on the project indicated that this was "natural" due to roundoff error, and it was observed earlier but not reported. Release of the program was stopped, followed by extensive tests, which indicated that the developed system was wrong.

> Extensive system verification should be carried out by "experts" outside of the development team prior to release of any ES. In-house testing of ES, regardless of the possible high degree of expertise, is not sufficient.

Three case studies illustrate the effects of the ramifications of the "incorrect tests" on the software developers. In one case the software was correct, to the best knowledge of the developers, however, it failed to simulate a reported laboratory test. After months of efforts to find the bug(s), the thrust of the inquiry was changed. A persistent interaction with the experimenters revealed that during the testing of the component, the specimen had slipped. This had substantially changed the support conditions, but was not reported in the publications. By simulating this slippage in the analytical model, results similar to the one reported by the experimenters were obtained.

Results emanating from a scaled down laboratory testing could not have been simulated via the analytical model. Again the analytical model versus test results showed unacceptable differences for certain response characteristics. This time, rather than initially questioning the software, the reported tests were subjected to close scrutiny. It was noted that due to scaling down of the physical specimen some alterations were made in the laboratory model. The experimenters assumed that these were too trivial to be of concern, thus they were not reported. Again, changing the analytical model to comply with the laboratory experiment, rather than the original structure revealed a very good correlation.

In the above studies highly reputable engineers had suggested that the analytical model may be good, but not good enough; thus, a correction or a correlation factor should be used to obtain the correct results. These experts have had very extensive field and practical experience, but their experience with computer-based engineering "simulation" was limited.

> No calibration factors to "improve," or to force the results should be embedded in ES. If necessary, such factors can be suggested by the ES, in addition to the original response of the system.

Internal Checking

ES is usually developed in such a manner that the input required for the system is minimal. However, if the number input values are somehow increased, it would than be possible to develop simplified internal checks to catch any inconsistency of the data. For example, in the handling of the bridge structures via discrete analysis scheme, e.g. finite element method, both the span length and the length of the longitudinal elements can be required. The sum of the longitudinal length of the elements should be equal to the span length, otherwise the program should reject the "data," (e.g. Ref. 5). However, if the program will ask only the span length of the bridge and will perform internal discretization, the opportunity for this audit is missed. Again, as related to the bridges, if the user inputs a 500 foot span length for a bridge with AASHTO Type-I prestressed concrete I-beams, the ES should at least flag the user (a) of the excessive span length, and (b) inappropriate beam section for the given span length. However, the rejection of the "problem" may severely limit the system. For the dimensions given above the user may very well be conducting a parametric study, and the values inputted may very well correspond to the "asymptotic" values. The system must identify the "unreasonable values," and should have the capability to be overwritten by the user. In a situation like this a highly visible message must be included in the output.

A general formula developed by the author and his colleague permits the establishment of full stress-strain curve for concrete with very high degree of accuracy. This program requires only the compressive cylinder strength. The formula has been thoroughly tested for 1 ksi to about 8 ksi concrete. With the new chemical additives, it is possible to obtain concrete with much higher cylinder strength. In an ES system if the user inputs 250 psi or 15 ksi cylinder strength, the system must reject the data and/or the problem. These values are outside the verified range of the system.

> ES must have as many internal controls as possible to check both the accidental and the systematic input errors by the users. Depending upon the "acceptability" of the values in question the ES should either reject the problem, or should continue after being "vetoed" and overwritten by the user.

Conclusions

The conclusions drawn and the recommendations made are highlighted in the paper. It is recognized that not all of the recommendations can be literally implemented; however, in order to have a smooth transition from the development stage of ES to the production environment the noted observations should be seriously considered.

Acknowledgements

The author extends deep appreciation to colleagues involved in the development, testing, and implementation of various expert systems in various organizations. Without their interaction with the author, this paper could not have been written.

References

1) D'Ambrosio, B., "Expert Systems-Myth or Reality?," BYTE-A Small Systems Journal, pp.275-282, McGraw-Hill Publishing Co., January, 1985.

2) Fenves, S. J., "What Is An Expert System," Expert Systems in Civil Engineering , (C.N. Kostem and M.L. Maher, Eds.), American Society of Civil Engineers, New York, 1986.

3) Fenves, S. J., Maher, M. L., and Sriram, D., "Knowledge-Based Expert Systems in Civil Engineering," IABSE Journal J-29/85, pp.63-72, Zurich, Switzerland, 1985.

4) Hayes-Roth, F., Waterman, D. A., and Lenat, D. B. (Eds.), Building Expert Systems , Addison-Wesley Publishing Co., Inc., Reading, MA, 1983.

5) Kostem, C. N., "Design of An Expert System for the Rating of Multibeam Highway Bridges," Expert Systems in Civil Engineering , (C. N. Kostem and M. L. Maher, Eds.), American Society of Civil Engineers, New York, 1986.

6) Kostem, C. N., and Shephard, M. S., (Eds.), Computer-Aided Design in Civil Engineering , American Society of Civil Engineers, New York, 1984.

7) Ludvigsen, P. J., Grenney, W. J., Dyerson, D., and Ferrara, J. M., "Expert System Tools for Civil Engineering Applications," Expert Systems in Civil Engineering , (C. N. Kostem and M. L. Maher, Eds.), American Society of Civil Engineers, New York, 1986.

8) Rich, E., Artificial Intelligence , McGraw-Hill Book Co., New York, 1983.

9) Sell, P. S., Expert Systems - A Practical Approach , Halsted Press, John Wiley and Sons, New York, 1985.

10) Winston, P. H., Artificial Intelligence , Addison-Wesley Publishing Co., Reading, MA, 1977 (Second Edition, 1984).

11) "The Smarter Computer," Newsweek, pp.89-91B, December 3, 1984.

12) "Why Can't A Doctor Be More Like A Computer?," The Economist, pp.101-102, December 1, 1985.

Expert Systems in an Engineering-Construction Firm

Gavin A. Finn[1] (AM.ASCE)

Kenneth F. Reinschmidt[2] (M.ASCE)

Introduction

The potential role of expert systems in the engineering environment has, of late, been extensively publicized. Although it has been generally accepted that the field of engineering provides a wide variety of potential applications for expert systems, most developments have been academic in nature. Experience at Stone and Webster Engineering Corporation in the development of expert systems for engineering and construction applications has led to clearly defined priorities for implementation. A better understanding of the nature of these systems and their interaction with developers and end-users has been gained.

This paper introduces the technical aspects of expert systems, hardware implementation requirements, current software technologies, applicability of rule-based systems in engineering and construction, and specific systems developed at Stone and Webster Engineering Corporation (SWEC).

Definitions

Artificial Intelligence (AI): The subfield of computer science concerned with understanding human reasoning and thought processes, and the application of this understanding to the development of intelligent computer technologies. Branches of this discipline include (among others) robotics, vision, speech recognition, and expert systems.

---

[1]Systems Analyst, Stone and Webster Engineering Corporation, P.O. Box 2325, Boston, MA 02107.
[2]Vice President and Manager, Consulting Group, Stone and Webster Engineering Corporation, P.O. Box 2325, Boston, MA 02107

<u>Expert Systems</u> : Computer programs that perform specialized tasks requiring knowledge, experience, or expertise in some field. Perhaps a more descriptive term would be "knowledge-based-systems".

<u>Domain Knowledge</u> : The term 'domain' refers to the specific area of application, such as welding or vibration engineering. Domain knowledge is that knowledge which is specific to the domain, rather than general knowledge, or common sense knowledge.

<u>Inference Mechanism</u> : The software which controls the reasoning operations of the expert system. This is that part of the program which deals with making assertions, hypotheses, and conclusions. It is through the inference mechanism that the reasoning strategy (or method of solution) is controlled.

<u>Knowledge Base</u> : That part of the program in which the domain knowledge is stored, in the form of facts and heuristics.

<u>Knowledge Engineer</u> : A person who works with the expert(s) in order to provide the transition from human knowledge to the appropriate computer representation. Knowledge engineers provide the AI expertise, and play a major role in designing the system.

<u>Heuristics</u> : Expert "rules-of-thumb" that are usually empirical in nature, based on experience and intuition, with no mathematical or scientific proofs.

Expert Systems

Although the field of Artificial Intelligence has been active for more than twenty years, this area of research and application has only recently been heavily publicized. The attention paid to developments in this field is due to the realization of the potential application of some of these technologies in the government, industrial, and private sectors. Expert systems have been brought into the limelight by programs such as MYCIN, developed by the Heuristic Programming Project at Stanford University, and XCON (or R1), developed by Carnegie-Mellon University and Digital Equipment Corporation. MYCIN performs diagnoses of infectious diseases. XCON performs the detailed configuration of VAX computer systems and their peripherals, now on a routine basis. Proposals for the extension of this technology (developed through the 1970's and early 1980's) into other fields has been widely imaginative, and often idealistic. There are significant potential application areas within the capabilities of current technology, however, that could represent improvements in technologies, performance levels, and product reliability.

What are the Advantages to Expert Systems in Engineering and Construction Firms ?

Expert systems allow for a distribution of expertise, such that the user can gain access to knowledge and logical processes that are known to others (the experts). Through the distributed use of these programs, a greater degree of consistency can be achieved and maintained. Higher accuracy and performance levels will be attained due to a continuous availability of high-level knowledge.

Access to high-level expertise can be made available on a twenty-four hour a day, seven day a week basis. This reduces overall cost and eliminates time-delays incurred as a result of the expert's prior or alternate commitments. Immediate access to expert knowledge can reduce down-times for machinery and non-productive time for labor. By means of the use of heuristics (possibly in conjunction with, or as enhancements to algorithmic procedures) quick and inexpensive solutions may be obtained without the need for complex and detailed simulations and/or calculations.

Incremental growth and improvement of the systems can be achieved through the experience of multiple users. The capability to incorporate feedback from the users can be used to improve the performance of the programs.

The Structure of Expert Systems

As was mentioned earlier, there are two primary and distinct parts to these programs, namely the inference mechanism and the knowledge base.

The Knowledge Base

Although there are numerous theoretical approaches to representing knowledge, most of these are highly complex and (as yet) impractical for commercial applications. In the industrial environment, the measure of applicability of a representation method must be its ability to accurately represent expert knowledge in a form that is clear and accessible to both the expert and the inference mechanism. Most of the existing expert systems utilize a knowledge representation method that is based on production rules. While the rule-based approach may seem simplistic, it can be a highly successful method of representing knowledge that is not algorithmic or deterministic.

Rules consist of a set of conditions, and a set of actions. If all of the conditions in a rule are true, then the actions are executed. The conditions are contained in the "IF" part of the rule, and the actions in the "THEN" part, thusly :

```
IF      condition_1
   AND  condition_2

THEN    action_1
   AND  action_2
```

Because human experts are not always certain about these rules, the conditions and/or actions may have certainty factors associated with them. This allows for a more realistic representation of the expert's knowledge, based on his degree of confidence in each rule. It is not always clear, however, that a certainty factor can be assigned in absolute terms. Not only is it difficult for a single expert to be consistent in assigning certainty factors, but there is usually a great deal of conflict when multiple experts offer their opinions. Consequently, it has been found that the assignment of certainty factors should be relative, indicating the relationship of one association to another, rather than the absolute probability of any given association.

Another interesting problem associated with certainty factors is the combination of already assigned values, based on new knowledge. If an assertion has been made, and the same assertion is an action of a second rule, then the existing certainty factor must be combined, in some manner, with the new one. Many methods are used to perform this combination. For example, some programs average the two values! So, if it has already been asserted that "action_1" is a possibility, with certainty factor of 50/100, and a new rule indicates that action_1 is possible with certainty factor 25/100, then the new associated certainty factor for action_1 would be 37/100, in between the values for the two rules..

In another method, a modified probabilistic formula is used to calculate the new certainty factor. This formula reads:
$$CF = (1-((1-CF_1)(1-CF_2)))$$
(This approach is based on the assumptions that the confidence factors are absolute probabilities that each assertion is true, and that the rules are independent. The formula expresses the probability that one or both of the rules gives a true answer, which is one minus the probability that both are false.)

So, using our previous example, applying this formula would yield the certainty factor of 63/100. Note how different this result is from the previous method's result. In one case the certainty increased, and in another, the certainty decreased!

There is no theoretically correct method for combining confidence factors, as there is for computing probabilities. The second method shown above has more support than the first, as it seems reasonable to believe that if a conclusion is reached by two paths, then the confidence in its validity should be greater than if it had been reached by only one path. Therefore the averaging method may not seem appropriate.

Lacking a definitive theory, however, it seems reasonable to provide several approaches, and to let the system developer choose that which he prefers (as only he knows the subjective meaning of the confidence factors !).

The Inference Mechanism

There are two basic forms of computer-based reasoning, these being forward-chaining and backward-chaining. In both methods, the program acquires information either in the form of a question to the user, or by means of accessing other programs and data bases.

In backward-chaining systems, the program has a built-in method for making an initial hypothesis as to what the solution is. That is, it assumes one possible solution to be true. The procedure then attempts to prove that the assumption is correct, by asking the user (or using its own inference capabilities) to confirm all of the prerequisite conditions for this particular solution. If the solution is disproved, through the non-existence of a prerequisite condition, for example, then the program chooses a different possible solution, and proceeds to prove this one in the same manner. Because this approach is based on assuming that the goal is known, backward-chaining systems are also known as "goal-driven" systems.

In a forward-chaining system, the program has no a-priori knowledge of the possible solutions. It uses the acquired information to evaluate the tree of possibilities, as it progresses through the solution procedure. Information is gathered, until the list of possible causes of problems has been narrowed down as far as is possible. This method of reasoning (or inference) is sometimes referred to as "data-driven", because it starts from the given data rather than from the assumed hypotheses. The reasoning progresses from an initial state (at which the program has no knowledge of the solotion), through intermediate states (in which the program's knowledge of the solution improves), to a final state (when the program has reached its goal.)

Application of Theory to the Development of Industrial Systems

For many years, organizations such as the National Science Foundation (NSF) and the Defense Advanced Research Projects Agency (DARPA) have funded academic research in the field of Artificial Intelligence. Recently, this technology has been successfully transferred to the industrial marketplace. Hardware and software products specific to the expert systems development environment have recently been produced, and are commercially available.

To develop an expert system application, SWEC defines a functional specification for the application before selecting a system configuration. The application is then analyzed and appropriate hardware and software tools are chosen for development of a prototype. Experience gained with the prototype then leads to the development of the operational product, perhaps using a different inference mechanism. In this manner, the operation of the system is not constrained by the incompatabilities between problem requirements and software capabilities, as would be the case if one tool was used for all applications. In selecting an appropriate tool, it is important to ensure that all of the important knowledge can be completely represented, and that the program's reasoning process allows for adequate representation of the expert's solution strategy.

The development process can be summarized by the following nine steps:

1) Preliminary Functional Specification
2) Selection of Prototype Tools
3) Prototype Development
4) Evaluation of Prototype
5) Revised Performance Specification
6) Selection of Operational Tool
7) Development of Operational Product
8) Feedback from Users (alpha- and beta-tests)
9) Enhancements (GO TO STEP #5)

Programming Languages and Tools

In academic research, expert systems have almost invariably been written in a computer language called LISP, developed at MIT in the early 1960's, although the Japanese and Europeans favor a language called PROLOG. Few programmers outside of the AI research environment are proficient in these languages, however. Consequently, inference mechanisms have been developed and offered for sale. The intent of these programs is to eliminate the need for knowledge engineers or experts to utilize the AI languages directly, in order to accelerate the development and use of expert systems in practice.

The accepted practice in the AI community is to use the specialized languages exclusively. In engineering practice, however, there are other priorities to be considered. While LISP and PROLOG offer definite advantages over procedural languages (such as FORTRAN and BASIC) with regard to symbolic processing, it is not necessarily expedient to use expert systems developed in these languages. Neither LISP nor PROLOG are easily compatible with other languages, so that existing programs (written in FORTRAN or BASIC) would most likely have to be rewritten, if they are to be used in conjunction with the expert systems. This would require major programming efforts, and significant expenditure of other resources (capital and labor). In developing expert systems for engineering applications it is highly desirable to make use of the existing base of knowledge (in the form of algorithms, subroutines and programs), in addition to rules. This engineering approach may differ from some other application areas, where there may not be an existing usable program base.

The alternative to utilizing AI languages for system development is to use the commercially available inference mechanisms. The level of sophistication of these programs ranges as widely as do their costs. Table-1 illustrates a representative sampling of the currently available systems and their hardware requirements. Some of these programs have the capability of utilizing existing software (from in-house engineering subroutines to packages such as LOTUS-123) within the framework of their operation. This capacity allows for the development of engineering application systems that have the capacity to utilize heuristic expertise as well as scientific calculations.

There are a large variety of commercially available shells. Some programs are "example- based", allowing the developer to input cases, rather than general rules. The shell then generalizes from the set of examples to form its own rules. Other shells allow the developers to input the rules directly, in the IF..THEN format. Some programs require the use of LISP-like rules, while others allow the use of subsets of the English language for rule input. Many programs require the use of text files for the storage of rules, while others utilize specially formatted files, created in an interactive rule-building session between the developer and the shell.

PRODUCT	VENDOR	HARDWARE
ART	Inference Corp.	LMI, VAX
KEE	Intellicorp	SYMBOLICS 3600 Class/ TI Explorer
S1	Teknowledge	SYMBOLICS 3600 Class/ VAX 11/780
M1	Teknowledge	IBM-PC
RULE-MASTER	Radian Corp.	Apollo Class/ IBM-PC
EXPERT	IBM Corporation	IBM Mainframe
INSIGHT2	Level 5 Research	IBM-PC
PERSONAL CONSULTANT	Texas Instr.	TI-PC/ IBM-PC
EXSYS	EXSYS Inc.	IBM-PC
TIMM	General Research	IBM/ DEC/ PRIME
ES/P ADVISOR	Expert Software	IBM-PC
Expert-Ease	Expert Software	IBM-PC

Note : Included in the Symbolics 3600 Class are the XEROX AI machines, and the Lisp Machines Inc. (LMI) computers.

Table 1

Stone and Webster Engineering Corporation has developed its own inference mechanism, Microcomputer Artificial Intelligence Diagnostic Service, for use on the IBM-PC class machine. It is a forward-chaining rule-based program that uses a subset of the English language for representing the rules. The program has two modules, one which 'compiles' the rules from their English-like format into a lower level representation, and another module for 'executing' the rules. Two applications developed using MAIDS^{TM} (PumpPro^{TM} and the Unit Commitment Advisor), are discussed in the section on current applications.

Because each application offers its own set of needs and limitations, it is wise not to limit the development process to any one shell or inference method. In addition to MAIDS^{tm}, Stone and Webster uses a number of commercial packages for applications development.

Hardware Requirements

Many expert systems applications are developed for specialized AI computers, such as LISP machines. The most sophisticated inference mechanisms are generally geared toward workstation class computers, which have built-in LISP interpretation and execution capabilities. These machines generally provide a complete AI environment, for system development. Other computers are also being used for AI applications, including IBM mainframes, DEC VAX computers, Apollo and SUN computers, as well as microcomputers.

When developing engineering application systems, the identification of the intended group of end-users, and their familiarity with and accessibilty to different computers are major considerations. In some applications, it is feasible to invest in a stand-alone AI machine, if the application is intended for a single user at a single location. If the product is intended for distribution to a multi-user group, then the hardware cannot be so limiting.

Stone and Webster has standardized on IBM-PC microcomputers and has distributed a number of them throughout its offices and field sites. This availability provides an existing hardware vehicle for product delivery. Another consideration is that some of the experts have little or no experience with computers. It has been found that access to a personal computer affords an environment in which people are most willing to participate in the expert system development process. The availablity of MAIDS^{TM} and commercial packages for the IBM-PC has further encouraged microcomputer based systems development.

SWEC has also installed a VAX-based inference mechanism, and has transferred some of the PC-based systems to one of the Boston Headquarters VAX computers. A dial-up service allows users to access the SWEC expert systems from any PC with a modem.

Experience has shown that initial expectations by industry analysts and AI researchers, that the memory capacity and CPU power of microcomputers would preclude the successful development of micro-based systems, were incorrect. Many of the micro-based inference mechanisms provide great storage, access, and processing power utilizing standard PC equipment, while there are a great number of engineering and construction applications that do not require large-scale or specialized computers.

Industrial Applications

Because of the level of detail required in engineering expert systems, the development of general problem-solving programs is far more expensive and complex than the development of systems that solve narrowly-defined, specific problems. There is a tendency, in selecting expert systems applications, to be far more ambitious than necessary. In selecting applications, it is prudent to begin with relatively small problems. Experience has shown that the level of complexity of a given problem is far greater than most people expect, from a problem description and reasoning perspective. The consequences of attempting problems without a reasonably accurate measure of the extent of the problem and its solution procedures could include incomplete systems, major development cost overruns, project abandonment, and even disillusionment with the entire field. There are no standards or guidelines for the assessment of a problem's size and complexity. It has been found, however, that a smaller problem domain can be expanded easily, whereas larger problem specifications are more difficult to prune. It is SWEC policy, therefore, to define the problem as specifically and as narrowly as possible, before any development work is begun. Inevitably, the problem's complexity grows naturally.

Applications in Power Systems Operations

Stone and Webster has developed a prototype expert system for power systems operations. The program provides advice to system dispatchers of electric power generators, combining the heuristic and economic/engineering optimization approaches. This demonstration system, the Unit Commitment Advisor, was developed using the MAIDS^{TM} inference mechanism, in combination with economic dispatch algorithmic subroutines.

In operation, the economic dispatch algorithm computes the optimum production schedule, determining the plants which should generate power hour-by-hour for the week ahead, to minimize operating costs. This optimum plan may not be realizable or desirable due to practical considerations. The expert system then processes the week's startup list, in terms of practical factors that the algorithmic method cannot include. The influence of weather forecasts, take-or-pay fuel commitments, maintenance schedules, operators' experience, and practical system requirements are some of the considerations that are included in the reasoning process. The system uses graphical displays to illustrate projected loads, required generation, and actual generation, in addition to a graphical breakdown of any hour's generation statistics. The user has the capacity to question the program regarding availability of units and explanations of the program's output.

This expert system is a prototype in the sense that actual operational programs must be tailored to the specific characteristics of each power system. The approach is applicable to many kinds of dispatching problems, and demonstrates that heuristic rules, optimization algorithms, and graphical displays can be successfully integrated in a PC-based expert system.

Applications in Construction, Operations, and Maintenance

Construction, operations, and maintenance are fertile fields for the application of expert systems, because so many of their functions are carried out at locations far from the home office, where experts may not be available at the right time. Problems are often unique, so that maintaining full-time experts at each site is not cost effective, and decisions must often be made immediately - there may not be time to bring in outside experts. Much of the expertise in these areas is based on experience and judgement, rather than theory and analysis, so that expert systems methodologies are appropriate for many of these problems.

A primary factor opening up construction and plant services to expert systems applications is the proliferation of microcomputers at these locations. Three or four years ago, it was rare to find a microcomputer at a construction site. Now, it is rare to find a construction site, even the smallest job, without at least one microcomputer. There are reported to be 41,000 microcomputers currently in use by the construction industry, and 85,000 in use by engineering firms. The computers at these sites are there to perform project planning, scheduling, cost accounting, payroll, purchasing, material control, work tracking, and many other functions. Their existence there offers unique opportunities for delivery of microcomputer-based expert systems.

Stone and Webster has recognized this opportunity, and has engaged in the development of expert systems for various aspects of construction and plant operations and maintenance services. Some of these systems are described below.

Centrifugal Pump Failure Diagnosis

There are large numbers of pumps in operation at most power and process plants, and reliable operation of these pumps is critical. In many cases, available maintenance personnel may not have the expertise to diagnose the causes of pump problems, even though they are charged with the task of keeping the plant in operation. Consequently, consultants may have to be called in to investigate poor performance or failures. The expert consultation can be costly and time-consuming, and the overall diagnose-repair process can thus be inefficient, particularly when the problem could be solved by on-site personnel if they could correctly diagnose the cause.

Stone and Webster has developed a centrifugal pump failure diagnosis system, called PumpPro^{TM}. (A failure is considered to be any operational problem, even if the pump is not taken out of service.) The intent of the program is to allow mechanics, technicians, and millwrights to avail themselves of expert knowledge without the necessity of calling in a human expert consultant. The program was developed using the MAIDS^{TM} inference mechanism by a Stone and Webster pump specialist and consulting engineer. PumpPro^{TM} has been distributed to over four hundred users, and is being used for diagnostic and training purposes.

The operation of the program is separated into four major phases:

1) Identify the Symptoms. This is accomplished by means of the MAIDS^{TM} user interface, which consists of text displays, and a question/answer input format. Figure 1 illustrates a typical question and associated text display.

2) Identify the Causes. The program uses its forward-chaining inference procedure to apply the heuristic rules to the observed symptoms in order to identify causes.

3) Provide tutorials. Realizing the possible confusion and lack of understanding that may be caused by use of some specialized engineering terminology, the program includes a series of optional tutorials, aimed at helping the user understand terminology and procedures. These tutorials are invoked at the user's request, so that users who are familiar with the terminology may proceed directly with the program. In this manner, different levels of user groups can be accommodated without compromising the efficiency or accuracy of the program's operation.

4) Suggest remedies. After probable causes have been identified, the program will instruct the user on appropriate remedial action. If the solution of the problem is beyond the user's capabilities, he will be advised to call in a technical specialist.

PumpPro^{TM} diagnoses problems by means of twenty-two possible symptom classes, and a summarized pump history. It allows the entry of multiple symptoms, and takes this information into account when reasoning about probable causes. Approximately

three hundred and fifty rules are used in the cause
identification phase, and there are seven major tutorials
available, with many minor tutorials in the cause
identification rules. There are a total of approximately
seventy rules dealing with appropriate remedial strategies and
actions. All told, the system contains more than four hundred
and sixty rules. An example rule, illustrating the MAIDS
English-like format, extracted from PumpPro, is shown in
Figure 2.

```
+------------------------------------------------------------+
|                                                            |
|         Stone and Webster Engineering Corporation          |
|                 P U M P   P R O  (w/monitor)               |
|  Microcomputer Artificial Intelligence Diagnostic Service  |
+------------------------------------------------------------+
WHICH OF THE FOLLOWING DESCRIBE THE PUMP CAPACITY

  1 . PUMP CAPACITY IS ZERO

  2 . PUMP CAPACITY IS INADEQUATE

  3 . PUMP CAPACITY IS ADEQUATE


ENTER THE NUMBER CORRESPONDING TO YOUR CHOICE ? 3
```

Figure-1

```
      BEGIN RULE
CATEGORY    : 16
AUTHOR      : T.J.FRITSCH
DATE        : 3-29-1985
REASON      : EMPIRICAL
CONDITIONS  : PUMPED LIQUID IS CLEAN
ACTIONS     : CLEAR SCREEN
                DISPLAY BLOCK TEXT
                  CHECK SHAFT SLEEVES AT PACKING
                END BLOCK TEXT
              ASK IS SHAFT/SHAFT SLEEVE WORN
      END RULE
```

Figure-2

Vibration Analysis Interpretation

The process of diagnosing problems in rotating machinery is
dependent, to a large extent, on two factors: i) The data that
are required in order to make a diagnosis, and ii) The
expertise of the diagnostician in interpreting the data.
Vibration monitoring and analysis is used extensively in the
routine maintenance and trouble-shooting of most rotating
equipment. The process of data acquisition is well understood,
and the technology in this area is very highly developed.
Maintenance personnel routinely monitor, via permanent or
temporary sensors, the vibration levels of this machinery.
Engineers are then faced with the task of analysing the data,
and interpreting the analysis, in order to solve any existing
or potential problems.

It has been found that there are experts who can immediately
identify causes of excessive vibrations in rotating machinery
by means of a swift examination of a few vibration data. In
order to improve the performance level of the engineers who
are usually assigned the task of providing a diagnosis, expert
systems for providing advice regarding the probable causes of
vibration were developed. A system for the interpretation of
vibration analysis data for industrial fans has been
completed, and is in place at SWEC. These programs are
intended to aid the engineers, by providing quick and reliable
interpretations, and suggesting probable causes of vibration.

The systems were developed for the IBM-PC, and are configured
to run on a standard 128K, dual floppy drive machine. The
inference mechanism uses subroutines, developed at SWEC, for
the purpose of analyzing the output of a data collection
device, and for presenting graphic displays of the analysis
results. A VAX-based version has also been implemented, using
the inference mechanism installed on the SWEC VAX.

The programs are intended for engineers who have more than a
cursory familiarity with vibration engineering, so it is
assumed that vibration engineering terminology is understood.
The program operates in an interactive question/answer format,
obtaining most of the required information from the user. At
the user's discretion, the program can utilize the analysis
software for determining the predominant frequency ratios, or
it can obtain this information directly from the user. Figure
3 illustrates an example of the graphic display of analysis
data. Other questions relate to the specifics of the machine
under investigation (bearing type, for example), and other
optional measured data. After obtaining as much information as
it can, the program presents the user with a ranked list of
probable causes of vibration, and offers a brief explanation,
where necessary. The program will provide explanatory
information about the conclusions, if asked for. Currently,
the system has over one hundred rules and diagnoses eighteen
causes of excessive vibration individually, as well as all
combinations of causes.

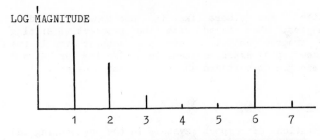

RATIO OF FREQUENCY TO RUNNING SPEED

Figure-3

Welding Programs

SWEC has developed an expert system to select the appropriate welding procedure, based on types of materials being welded, type of weld required, etc. The system selects the appropriate procedure, and then prints it for the welder to use in the field. The system operates on an IBM-PC, and is currently being used by field personnel on job sites. Since microcomputers are already available on these sites, the welders have immediate access to information, and the correct welding procedure is always selected.

An expert system for the diagnosis of field welding defects has also been developed. This program provides immediate advice to welders or supervisors, helping them determine the causes of field welding defects. The program takes into account different welding procedures, and code requirements, site conditions, observations, etc. Well over fifty causes of defects are dealt with, requiring the implementation of hundreds of rules.

In order to be approved for specific construction jobs, welders must have been qualified for the appropriate materials and procedures. The qualification procedure can involve combinations of tests, and may be quite complex. SWEC has developed an expert system which details the appropriate qualification tests for each job application, enabling managers to efficiently establish complete job qualification. Customized systems for clients and construction sites are developed.

Project and Construction Management

Project management software usually presents the user with large volumes of output, consisting of detailed reports of float times, activity lists, and other such data. In order for this information to be useful in the practice of project management, it must be interpreted, and related to a project status. Is the project running smoothly?

This- question and others like it are commonly asked by
project managers, when faced with the prospect of sifting
through the program output. Stone and Webster is in the
process of developing expert systems to bridge the gap between
the project management software and the project managers.

Conclusion

The implementation of expert systems in the engineering and
construction workplace provides significant benifits. Expert
systems for advising engineers and technicians are in
existence today, and are being effectively used to help solve
real-world problems. The availability of expert
system-building shells has greatly improved the technical
development capability within the engineering community, and
facilitates an ever-increasing effort to implement these
programs. Different approaches to the delivery of such
programs for engineering applications (including the use of
microcomputers and VAX-class machines) allows for a wide range
of distribution capabilities, thus broadening their overall
effectiveness .

Typical applications of engineering expert systems include
machine diagnostics, power system operations, welding
procedure selection, welding defect diagnosis, and project
management analysis. At Stone and Webster, these applications
are directed toward providing an interactive consultation
environment, such that users (at many different technical
levels of experience) can utilize high-level expertise in an
efficient manner.

In the long term, the influence of expert systems for
engineering will be seen in the enhancement of service
capabilities, and the improved productivity and efficiency of
engineering-construction firms.

Howsafe: A Microcomputer-Based Expert System To Evaluate the Safety of a Construction Firm

Raymond E. Levitt, Member ASCE[*]

Abstract

Expert systems are computer programs which use knowledge obtained from experienced practitioners (experts) in a specific domain to assist others (users) in solving complex problems at expert levels of competence. Hallmark characteristics of an expert system include flexibility for the expert to view and modify the knowledge that the system will use to reason about the problem, and the ability of the system to explain to a user why it needs certain data or how it has reached a given conclusion.

Expert systems have been especially successful for diagnostic applications. Example diagnostic expert systems include **Prospector,** a system for diagnosing the presence of mineral deposits, and **Mycin,** one of the earliest expert systems, which diagnoses infectious blood diseases. Virtually all previously developed expert systems described in the literature run on minicomputers or LISP workstations.

Howsafe, the expert system described in this paper is also a diagnostic expert system, but it differs from these prior efforts in two ways:

1. **Howsafe** is developed and runs on an **IBM Personal Computer** using **The Deciding Factor** [Campbell, 1985], an expert system development language or "expert system shell"; and

2. **Howsafe** deals with diagnosis of an organization's structure and operating procedures -- i.e., it diagnoses malfunctions in a **social** system, rather than in a mechanical or biological system.

This paper discusses the development of the **Howsafe** expert system as a case history in knowledge engineering on microcomputers. The focus is on issues of knowledge representation, reasoning and explanation which arose in the development of **Howsafe** rather than on the domain knowledge of construction safety management. In particular, the paper addresses the issue of knowledge engineering when the expert system developer or "knowledge engineer" is also the expert. We conclude that microcomputer expert system development tools like **The Deciding Factor** can help to resolve the current knowledge engineering bottleneck by permitting experts to carry out much of their own knowledge engineering.

---

[*] Associate Professor, Department of Civil Engineering Stanford University, Stanford, CA 94305

Background

Construction Safety Research.--Stanford's Construction Engineering and Management Program has been conducting research on construction safety since 1969 [Levitt, 1975; Hinze 1976; Samelson, 1977; Levitt et al, 1981]. Up to now, the knowledge on this subject that has been painstakingly gathered over the past 15 years has been disseminated in traditional ways: technical reports, journal articles and seminars. These traditional ways for university researchers to distribute knowledge gained from research have been reasonably successful in this case. Nevertheless, a manager who wants to use this knowledge in order to evaluate a project's or firm's safety practices must still read through volumes of technical reports to get to the recommendations, or must try to find journal articles which may not be in the company's home office library, much less in the jobsite trailer.

Recognizing the need for more convenient means of communicating this knowledge to field construction managers, in 1983 the author and a colleague began writing a handbook on the subject of construction safety [Levitt and Samelson, 1986]. At the same time, the author began to explore the feasibility of developing a series of microcomputer-based tutorials to evaluate safety-related aspects of a contractor's organization and procedures. The aim of these tutorials was for them to educate the user about construction safety management while they carried out their evaluations. To do this, it was felt that the programs should avoid being "black boxes" which accepted input, ground away, and delivered output, leaving the user with no more knowledge about how to carry out the diagnosis than when he or she began.

By 1984, many construction contractors had begun to adopt the IBM Personal Computer (PC) as a standard and were starting to install PC's in their field trailers. In order for the proposed tutorial systems to be useful to them, it was decided that these diagnostic programs would be developed to run on personal computers. The quest for a "glass box" rather than black box computing environment that would run on personal computers led the author to knowledge based expert systems.

Evolution of Knowledge Based Expert Systems.--During the late 1970's and early 1980's, the first knowledge based expert systems were moving from university computer science departments and research laboratories into commercial and industrial use. The MYCIN expert system was reported as diagnosing blood diseases at the level of medical interns [Shortliffe, 1976] and the Prospector expert system was reported as having identified a large molybdenum deposit [Campbell, 1982].

After earlier attempts to incorporate wide-ranging expertise into computer programs had stalled, these expert systems had been purposely designed to be very narrow and specific in their focus. For example, MYCIN was able to diagnose infectious blood diseases but had no capability to carry out any other type of medical diagnosis; and Prospector was really a series of expert systems, each of which dealt with a particular type of mineral deposit.

In the late 1970's, the developers of the MYCIN medical diagnosis expert system at Stanford reported on E-MYCIN (Essential MYCIN) as the first expert system "development shell" [van Melle, 1981]. E-MYCIN consisted of a generalized version of the control structure and user interface from MYCIN, with the infectious blood disease knowledge stripped out of it. This Essential-MYCIN or Empty-MYCIN permitted a "knowledge engineer" to insert other medical knowledge, or knowledge from a non-medical domain, readily into it. The developers of Prospector subsequently announced the **Knowledge Acquisition System (KAS)**, a generalized expert system shell based upon the control structure of **Prospector**. Others followed.

One of the first non-medical applications to be developed using **E-MYCIN** was **SACON**, an expert system to assist structural engineers in configuring difficult structural analysis problems for processing by a finite element package [Bennett et al, 1978]. This was getting closer to the construction safety application area, and the demonstrated generality of the tools was exciting, but the computing environment was still mainframes or minicomputers.

During 1983 the author learned from Dr. Alan Campbell, one of the scientists from the **Prospector** project, that he was working on a microcomputer expert system development shell based upon his experience with **Prospector**. Campbell felt that he could produce an IBM PC expert system shell that would have all of the features of **Prospector**, but which would be so easy to use that even a novice computer user could easily model his or her own, or an expert's, knowledge. And he said that he planned to sell it for under $200 so that it could be as widely used as spreadsheets or word processors!

The author volunteered to Beta-test this new program on the **Howsafe** application and, in early 1984, received a prototype of **The Deciding Factor** to test. The structure of **Deciding Factor** exactly matched the needs of the **Howsafe** application, since it was specifically designed for developing diagnostic expert system applications. The result was rapid development of the **Howsafe** expert system tutorial and a smooth interface for both expert and user.

Knowledge Representation in Howsafe

What is the structure of the knowledge that an expert uses in diagnosis and how should it be represented in a computer program? This question was addressed by the developers of **MYCIN**, **Prospector** and other diagnostic systems. We share their views about the structure of diagnostic knowledge, but have used a somewhat different style of representing it.

<u>Some Previous Approaches to Modeling Diagnostic Decisions.</u>--The consensus among researchers who have worked at modeling diagnostic decisions seems to be that in carrying out a diagnosis, an experienced practitioner attempts to infer the degree of truth or the strength of belief for a series of goals representing possible system malfunctions or other high level inferences, from a range of available evidence. There can be a many-to-many relationship between pieces of evidence and goals inferred from them, and the degree of belief in a piece of evidence can range from positively true to positively false.

In the medical domain, this translates into establishing a series of pathologies as goals whose truth must be determined from consideration of symptoms reported by the patient or from laboratory test data. In the mineral geological domain, the top level goal is the existence of a given type of mineral deposit. Intermediate level goals used to infer this goal are a series of regional and local geomorphological, geophysical or geochemical patterns based upon ore genesis theories or empirical correlations. The presence of these patterns is inferred from surficial geological observations, from interpretation of geophysical test data, or from analysis of samples.

In medical diagnosis systems like **MYCIN**, following completion of diagnosis, prescriptions can be tailored to each patient's needs, taking into account factors such as patient age and weight, and drug interaction effects. In the **Prospector** situation, once a deposit has been identified, the decision by a mining company of whether the deposit should be staked, developed, sold or abandoned involves a complex interplay between organizational needs, macroeconomic and microeconomic trends and human factors. Consequently **Prospector** did not attempt to provide prescriptions. It stopped at establishing the strength of belief that a mineral deposit of a given type was present in a particular region, and displayed this as a color scaled map.

Modeling the Howsafe Diagnosis.--For **Howsafe**, we have adopted the latter

approach. The program diagnoses strengths and weaknesses in a construction firm's organization and procedures which could impact safety performance. The style of doing this conveys to the user the underlying causes of strengths or weaknesses, and leaves implicit the strategies that should be taken to improve the organization's safety performance -- i.e., it is implied that the user should concentrate on eliminating the identified causes of weaknesses and on maintaining existing strengths.

The knowledge to be represented in **Howsafe** starts out with a top-level hypothesis whose strength of belief the user seeks to determine, "This construction firm has the required organization and procedures to promote safe construction." This hypothesis is inferred from a series of intermediate goals such as "Top management truly cares about safety," "Managers at each level are held accountable for the safety of all of their subordinates," etc.

Each of these intermediate goals is then itself treated as an hypothesis with lower level evidence to determine its truth value. For example, whether "Top management truly cares about safety" is determined from the most strongly believed of the following: "Top management knows all workers and their families personally," or "Clients weigh the company's safety record as a factor in negotiating contracts," etc.

The **Deciding Factor** system is designed to work in exactly this way. Thus the system developer (either the expert or a "knowledge engineer" working with an expert) structures the knowledge like an inverted tree, with the top-level diagnosis on the top, supported by lower level inferences, pyramiding down three or four levels to **factual assertions**, whose validity can be objectively evaluated by the user, at the bottom end of each branch.

Each leaf node of this inverted tree or "inference net" corresponds to a question which the user will be asked in its automatically parsed form: "To what degree do you believe that ...**assertion**... ?" or in the form of a customized question crafted by the developer. For example the **Howsafe** program will ask a user the customized question, "Are records of accidents kept which permit senior mangers to identify the number of injuries to each superintendent's workers?" The user answers all questions on a scale from +5 (positively true) through 0 (don't know) to -5 (positively false).

This approach to structuring knowledge is functionally equivalent to a production rule system with certainty factors in which rules must be organized hierarchically. Although the requirement to organize all rules hierarchically may appear to be a restriction, anyone who has built an expert system with more than 50 rules will verify that production rule systems which permit rules to be listed in random order are difficult to create -- and even more difficult to modify -- without introducing hanging rules or other logical errors.

Based upon experience with several different types of production rule systems, the author asserts that the requirement for hierarchical organization of rules in "problem decomposition" style expert system frameworks like **Prospector** and **The Deciding Factor** is a valuable discipline, and an aid to knowledge engineering. Moreover, as explained below, the possibility to repeat identical evidence or hypothesis strings in different parts of the inference net permits the knowledge engineer to get around the restriction of organizing all ideas in a strict hierarchy.

A portion of the inference net for **Howsafe** is shown in Figure 1.

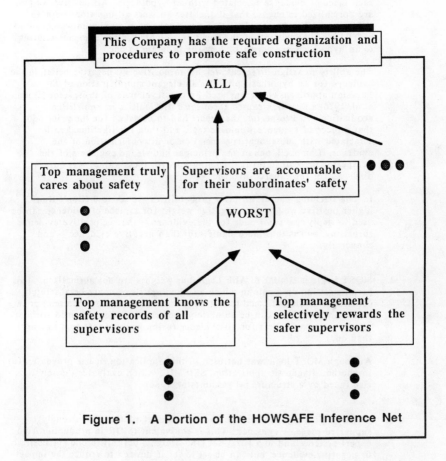

Figure 1.   A Portion of the HOWSAFE Inference Net

Reasoning in Howsafe

We have explained how the ideas in **Howsafe** are organized hierarchically as a series of evidence ==> hypothesis relationships. In this section we explain how the system infers belief in an hypothesis from the user's expressed belief in the pieces of evidence that support it.

The Deciding Factor provides four different ways to combine beliefs for evidence statements associated with a single hypothesis:

1. **ALL Logic** acts like a weighted sum, where the belief of each piece of evidence adds some weight to the belief of the hypothesis. The expert assigns weights for positive and negative belief expressed by the user for each piece of evidence associated with an hypothesis. All positive weights are normalized to one, so that if the user answers all questions with +5 (**YES** or **TRUE**), then the hypothesis will take on a maximum weight of +5 or **TRUE**. Similarly, all of the negative weights for the evidence relating to an hypothesis are normalized by the program to one.

    The ability to assign different weights to positive vs. negative belief in the evidence for an hypothesis provides an elegant simplification of the Bayesian approach which was used by the developers of **Prospector** [Duda, et al, 1976] and which caused them some difficulties in knowledge acquisition. In **Prospector**, the expert had to be asked for the prior odds that a piece of evidence would be true, and then for likelihood ratios associated with subsequent confirmation or disconfirmation of the evidence. Campbell, one of the principal knowledge engineers of the **Prospector** project, states that coming up with these values was quite difficult for the experts involved.

    In **The Deciding Factor**, all prior probabilities are set to **0 (Don't Know)**. A higher positive weight than negative weight for a piece of evidence can be used to imply low prior odds for the evidence -- that is, positive evidence is surprising, and is thus more significant than negative evidence -- and conversely.

2. **MOST Logic** is similar to **ALL Logic** but weights are not normalized. This permits setting up a rule in which an hypothesis can never be absolutely confirmed or disconfirmed (using weights that sum to less than one) or in which an hypothesis can be absolutely confirmed or disconfirmed without requiring strong beliefs for all evidence (using weights that sum to more than one).

    Although MOST logic was not used in this application, it has proven useful in another diagnostic application, **SEISMIC**, which evaluates the seismic risk faced by a structure [Miyasoto et al, 1986].

3. **BEST Logic** assigns to the hypothesis the belief of the most strongly supported piece of evidence. This is equivalent to **OR** in production rule expert systems, and to a fuzzy set **OR**. Weights default to one for positive or negative evidence, but can be set lower if desired to reduce the impact of a piece of evidence.

4. **WORST Logic** assigns to the hypothesis the belief of the least strongly supported piece of evidence. This is equivalent to **AND** in production rule expert systems, and to a fuzzy set **AND**. As with Best Logic, weights default to one for positive or negative evidence unless set lower.

All of the above types of combination logic for evidence can be augmented by some additional programming features of **The Deciding Factor**. These features provide for branching and pruning of the inference net during a consultation.

1. **KILL Values** can be defined for any piece of evidence. These delimit an acceptable range for belief in the evidence, outside of which the hypothesis will be negated. A common use of kill values is to stop a line of questioning. This feature was called "context values" in **Prospector**.

2. **CONDITIONAL Logic** can be used with any of the four combination types listed above. It treats the first evidence for the hypothesis as a conditional question, testing its value against a high and low range as for kill values. If the response is in range, the conditional question is ignored and has acted only as a gate; the system proceeds to explore the remaining evidence. If the answer is out of range, then the belief for the hypothesis is determined by the conditional question alone, and the other evidence is not considered.

Both kill logic and conditional logic were employed extensively in **Howsafe**. These permit the system to be tailored so that the user's responses are sought only when needed, and consultations have an easy and logical flow. An attractive feature of **The Deciding Factor** is that it always permits a user to backtrack in a consultation and change a response previously entered. Thus, if a user finds that a response has eliminated an hypothesis, and wishes to go back and modify the answer, this can be done with ease.

The control structure provided by **The Deciding Factor** is depth first backward chaining. Starting from the top level hypothesis, the program attempts to satisfy the first goal at the next level **(The Deciding Factor** refers to these second level goals as "key ideas"). It then chains down, through the first piece of evidence listed at each level, to the bottom or "leaf nodes" of the tree which have no supporting evidence from which their belief can be inferred. The user is prompted to enter his or her belief for each "leaf node" idea that the system reaches.

The control structure follows this mode of depth first backchaining, with user prompts for unsupported ideas, except when it encounters evidence whose belief (entered or inferred) is outside of ranges delimited by kill values or conditional logic. These out of range beliefs cause the program to branch around, or prune, parts of the tree.

The ability of **The Deciding Factor** to interrupt depth first backchaining with kill values or conditional branching provides knowledge engineers with considerable flexibility at a relatively small price in training and ease of use.

The **Howsafe** application involved diagnosis of a social system rather than a mechanical or biological system, so it was necessary to permit users to give answers in the form of degrees of belief, rather than as responses to "true/false" or "choose an answer from a list" prompts. The ability to handle input in the form of degrees of belief is one of the strongest features of **The Deciding Factor** expert system shell. For applications where user responses are expected to be clear cut or discrete, other microcomputer-based expert system shells such as **The Insight Knowledge System** or M-1 may be more convenient to use.

The control structure of **The Deciding Factor** makes it possible to jump ahead to the conclusion at any point, and to jump back and forth between a particular piece of evidence and the conclusion in order to test the sensitivity of the conclusion to the particular piece of evidence.

When the system encounters a repeated idea for which it already has a belief, it assigns the repeated idea the belief previously entered or inferred. This feature provides the capability for non-hierarchical use of evidence, for generating interim summaries as dummy questions or hypotheses, or for pruning parts of the inference net, when used together with kill values or conditional logic.

The conclusion for a consultation is presented as the inferred degree of belief in the top level hypothesis, along with the weight of each of the next level subgoals or "key ideas".

The reliability of the conclusion (expressed as a percentage) is also presented. The author is unaware of other languages or shells which provide this feature. Reliability refers to the strength of user responses. If the user answers near +5 or -5 to most questions, this results in a high reliability score for the consultation; answers near 0 ("don't know) reduce the reliability. By default, each question has the same weight in determining the overall reliability of the consultation, but this can be overridden by the knowledge engineer. If the expert feels that a user should be certain about the answers to particular questions for a consultation to be "reliable", then these questions can be assigned a higher weight in determining reliability, and conversely.

Explanations in Howsafe

By combining the expert's knowledge represented in the inference net with its own natural language capability, **The Deciding Factor** can parse together intelligent, natural language explanations of its reasoning paths. It can also provide additional background information (including simple pictures) entered by the knowledge engineer to elaborate on terminology used in a question.

HOW.--As an hypothesis is being introduced, the user can ask how it will be evaluated. **The Deciding Factor** provides a "lookahead" and lists the evidence that will be used to evaluate the hypothesis, and the logic with which it will be combined -- ALL, BEST, WORST or MOST. Similarly, after an hypothesis has been evaluated the user can ask how the belief was inferred. The program provides a natural language "lookback", recapping the logic used and the weights of evidence considered.

WHY.--To get the context of a given question, the user can ask why it is being asked. The program provides a diagram showing the top level hypothesis, the hypothesis currently being considered, and all of the evidence for it. The question currently being asked is highlighted in this list, and the range of allowable answers that will not trigger a kill value is provided.

EXPANSION.--The **Howsafe** expert system contains numerous screens which provide additional background information to expand upon the terms in a given question. For example users not familiar with the meaning of "experience modification rating" can call up a customized explanation screen which explains the workers compansation insurance system and how experience modification rating is carried out. These screens are created in **The Deciding Factor** using its dedicated word processor. The knowledge engineer can create these screens using both text and graphical symbols from the extended ASCI character set, and then attach them to evidence or hypothesis statements.

The combination of automatically parsed natural language explanations and custom, expert-generated explanations provides an extremely friendly and informative run time computing environment for novice users of **Howsafe**. As a result, one can view expert systems of this type as a powerful new communication medium for knowledge about how to make decisions.

Issues in Knowledge Acquisition

The process of formalizing knowledge so that it can be incorporated in an expert system is often referred to as knowledge engineering. Many authors view the scarcity of knowledge engineering skills as the major bottleneck to more widespread application of expert systems [Hayes-Roth et al, 1983].

Until recently, the mechanics of interacting with the available expert system programming languages required considerable programming expertise, usually in one of the dialects of the LISP language, as well as long periods of training. Thus the tools themselves were part of the bottleneck.

The Deciding Factor and other microcomputer-based expert system languages such as **INSIGHT, EXSYS,** or **M-1** reduce the need for the knowledge engineer to be a computer scientist. Experienced users of microcomputer spreadsheets and databases have been able to make the transition to these expert system languages easily. The author has taught two day short courses in which persons with almost no prior computer background have been able to build simple prototype expert systems using **The Deciding Factor** with only limited assistance in selecting the application, and in creating the initial structure of the system. Thus the tools themselves no longer present a barrier to widespread use.

However, formalizing the knowledge to put into an expert system is at least as difficult as formalizing the knowledge to put into a good journal paper or book. It takes careful thinking and iteration. And authors' skills improve with practice. Identifying the best tool for a given application and structuring at least the top level ideas in an expert system take considerable practice to do well.

In the case of **Howsafe**, the author (as expert) participated in one initial session of about two hours in which Dr. Alan Campbell (author of **The Deciding Factor**) acted as the knowledge engineer. Once the initial structure of the expert system had been created through discussion and dialogue with an experienced knowledge engineer, the author was able to carry out the remainder of the knowledge engineering alone.

A similar path is being followed in developing another microcomputer expert system, **Conadmin**, which selects the type of contract to use for a planned construction project. In this case, the author initially acted as knowledge engineer for an expert, Mr. Donald S. Barrie, to help define the structure of the system, and the expert is currently proceeding with the detailed knowledge engineering and refinement of **Conadmin**.

In both of these cases, the knowledge engineer provided a sounding board for ideas, and guidance in structuring the expert's knowledge initially, but was able to withdraw gracefully after one or two sessions of a few hours each, and to provide critique and suggestions in a review mode thereafter. The expert was able to function as knowledge engineer for the majority of the detailed engineering. This model of experienced knowledge engineer and novice knowledge engineer (expert) is not unlike the manner in which an experienced structural or geotechnical engineer might collaborate with and guide a junior colleague. These experiences indicate that a tool such as **The Deciding Factor** permits a redistribution of the established roles between knowledge engineer and expert.

It is recognized that **The Deciding Factor** and similar tools are simple rule systems and do not offer the power of frames, graphics and multiple reasoning modes provided by the most powerful artificial intelligence programming environments such as IntelliCorp's **Knowledge Engineering Environment (KEE)** or Inference Corporation's **Automated Reasoning Technique (ART)**. Nevertheless, they are perfectly adequate tools for many simple applications, especially ones involving diagnosis.

Personal computer tools such as **The Deciding Factor** can also be used to prototype selected portions of a knowledge base that will ultimately be incorporated in a larger system using a development language such as **KEE** or **ART** on LISP workstations. This was done for the geotechnical rules in **Platform**, a project scheduling advisor developed in **KEE** [Levitt and Kunz, 1985]. This decentralization of some parts of the knowledge engineering task facilitates parallel prototyping of separate parts of the system, and can get experts involved in a more productive way in the development of expert systems.

Conclusions

After final validation, a commercial version **of Howsafe** will be released during 1986 along with a companion package **Safequal**. **Safequal**, also developed using **The Deciding Factor**, helps a construction buyer to select contractors based upon their past safety performance and current safety management practices. These expert systems, and others that are being developed, will convey knowledgable assistance to decision makers by acting as intelligent, interactive checklists or books. They truly represent a new communication medium for knowledge about decision making.

The development of **Howsafe** was completed in a fraction of the time (five man months vs. several man years) that an expert system of comparable depth and breadth would have required before the emergence of powerful and convenient expert system programming languages for microcomputers. **The Deciding Factor** was an excellent tool for development of this diagnostic application and provides a suitable run time environment for both advanced users and novices -- in terms of computer use, as well as construction safety.

Experience on this project and others indicates that, when using this type of programming environment, the knowledge engineer can delegate to the expert much of his or her traditional responsibility in detailing the knowledge base of an expert system.

The low cost, ease of use, power and flexibility of currently available microcomputer-based expert system tools can thus make a significant contribution towards eliminating the knowledge engineering bottleneck to widespread application of expert systems in Civil Engineering.

References

Bennett, J.S., L. Creary, R.L. Engelmore and R. Melosh, "SACON: A Knowledge Based Consultant for Structural Analysis" *Report No. HPP-78-23*, Computer Science Department, Stanford University, 1978.

Campbell, A.N., V.F. Hollister, R.O. Duda and P.E. Hart, "Recognition of a Hidden Mineral Deposit by an Artificial Intelligence Program," *Science Vol 217, No. 3*, September, 1982.

Campbell, A.N. and S. Fitzgerrell, *The Deciding Factor User's Manual*, Power Up Software, San Mateo CA, 1985.

Duda R.O., P.E. Hart and N.J. Nilsson, "Subjective Bayesian Methods for Rule-Based Inference Systems," *Proceedings of the AFIPS National Computer Conference, Vol. 45*, pp 1075-1082, New York, 1976.

Hayes-Roth, F., D.A. Waterman and D.B. Lenat, *Building Expert Systems*, Addison Wesley Publishing Co., Reading, Mass., 1983.

Hinze, J. "The Effect of Middle Management on Safety in Construction," *Technical Report No. 209*, Dept. of Civil Engineering, Stanford University, 1976.

Levitt, Raymond E., "The Effect of Top Management on Safety in Construction," *Technical Report No. 196*, Dept. of Civil Engineering, Stanford University, 1975.

Levitt, R.E., N.M. Samelson and H.W. Parker, "Improving Construction Safety Performance: The User's Role" *Technical Report No. 260*, Dept. of Civil Engineering, Stanford University, 1981.

Levitt, R.E., and J.C. Kunz, "Using Knowledge of Construction and Project Management for Automated Schedule Updating" *Project Management Journal, Vol. 16, No. 5*, December, 1985.

Levitt, R.E. and N.M. Samelson, *Construction Safety Management*, McGraw-Hill Book Company, in press, 1986.

Miyasoto, G., W. M. Dong, R.E. Levitt and A. C. Boissonade, "Implementation of a Knowledge Based Seismic Risk Evaluation System on Microcomputers" *Journal of Artificial Intelligence in Engineering, Vol. 1, No. 1*, 1986.

Samelson, N., "The Effect of Foremen on Safety in Construction," *Technical Report No. 219*, Dept. of Civil Engineering, Stanford University, 1977.

Shortliffe, E.H., *Computer Based Medical Consultations: MYCIN*, American Elsevier, NY, 1976.

Van Melle, W., *System Aids in Constructing Consultation Programs*, UMI Research Press, Ann Arbor, MI, 1981.

AN EXPERT SYSTEM FOR CONSTRUCTION SCHEDULE ANALYSIS

by

Michael J. O'Connor[1] M. ASCE
Jesus M. De La Garza[2]
C. William Ibbs Jr.[3] M. ASCE

Abstract

The objective of this project is to develop an expert system for the analysis and evaluation of construction scheduling networks from an owners' perspective.

The knowledge base, which is built from the input of several expert field engineers, consists of scheduling decision rules and construction knowledge. Scheduling decision rules include criteria for evaluating a contractor's initial planning schedule, i.e. the inclusion of procurement lead time for special purpose equipment, as well as rules for evaluating construction progress based on comparison with historical data and extrapolation of trends. General construction experience, such as the effects of weather, crew sizes, and reasonable placement rates, is also incorporated in the construction knowledge base.

Data from a relational database management system, user supplied project-specific information, and the knowledge base supply the necessary input to a microcomputer based expert system shell. The database environment primarily consists of output from a commercially available microcomputer based project management system, and other information such as productivity rates, external factors, and statistics related to the project under consideration.

Ongoing validation of this prototype confirms its ability to give results similar to those of a human expert.

[1] Research Team Leader, US Army Construction Engineering Research Laboratory, P.O. Box 4005, Champaign, Il 61820.

[2] Research Assistant, US Army CERL; Ph.D. candidate, Department of Civil Engineering, University of Illinois at Urbana-Champaign, Il 61801.

[3] Associate Professor, Department of Civil Engineering, University of Illinois at Urbana-Champaign, Il 61801.

Introduction

Integrated project management expertise about design, construction, economic, social, and political issues plays an increasingly important role in the proper execution of construction projects. Although computers have become indispensable tools in many endeavors, they have been mainly used for processing information that fits rigid structures and for generating large numbers of flexible (i.e. user defined) reports. We continue to rely heavily on the human's expert ability to identify, analyze, and synthesize diverse factors, to form judgments, evaluate alternatives, and make decisions. In sum, to apply his or her years of experience to the problem at hand. This is especially true regarding construction project management in which experience and subjective judgment play a major role.

Years of research in the field of artificial intelligence have produced effective expert systems for representing empirical judgmental knowledge and using this knowledge in performing plausible reasoning. Frames, rules, and logic are the three basic architectures for building such expert systems [2, 3, 4]. A frame-based expert system allows the developer to organize the knowledge base into a hierarchical taxonomy of frames. The inheritance mechanism built into these systems provides the main means for inferring. In a rule-based system, the knowledge base consists of self contained pieces of knowledge in the form of "if ... then" rules. Rule-based systems are characterized by forward and backward chaining inference mechanisms. Finally, the knowledge base in logic-based systems consists of a set of objects and predicates. Quantifiers and logical connectives are applied to these objects and predicates to form relationships. The theorem-prover inference mechanism insures that all and only valid consequences are asserted.

The following is a sample list of expert systems Buchanan [1] has reported as being either in routine use or field testing:

System:	Company:	Purpose:
XCON, XSITE, XSEL	DEC	Configuring and checking orders for VAX computers.
	Babcock & Wilcox	Advising on types of welds and materials based on engineering specs.
SECOFOR	Elf-Aquitaine	Advising on drilling.

System:	Company:	Purpose:
	IBM	Advising on the cost of moving mainframes from site to site.
Wheat Counsellor	Imperial Chemicals	Advising on the control of disease in winter wheat crops.
PUFF	Pacific Medical Center	Interpreting pulmonary function tests.
Dipmeter Advisor	Schlumberger	Analyzing oil well logging data.

While some of these systems are based on a single architecture, others are based on a combination of them. This research project and paper concerns a similar expert system for construction schedule analysis developed to help project managers analyze and evaluate initial as well as progress construction scheduling networks. The system combines both rules and frames architectures.

Overview of the Construction Schedule Analysis System

The construction schedule control domain is characterized by the use of expert knowledge, judgment and experience. As generally practiced, construction schedule control does not begin to exploit the state-of-the-art of network analysis systems. Traditionally, less sophisticated owners tend to limit their analysis of the contractor's network to such parameters as 1) Anticipated project start date; 2) Estimated project finish date; 3) Total project duration; and 4) Total project cost. There is a presupposition in this approach to contracting that does not always hold true, that contractors thoroughly verify their project schedules. If an expert system can provide the expertise that knowledgeable project managers use to evaluate schedules, then owners and other individuals can manage their construction projects more effectively.

This Construction Schedule Analysis system is intended to emulate the reasoning process of experienced project managers in assessing the correctness and soundness of a given project schedule. Nevertheless, to intelligently judge the validity of a proposed schedule, the system user must know the scope and details of the work and the requirements affecting it. At no time will the system relieve the user of the responsibility of independently insuring that the schedule is correct; rather it will detect difficulties with some degree of reliability and propose potential reasons for scheduling problems. To do this, the knowledge base contained in the system consists

of a set of established scheduling decision criteria [8]
and construction knowledge. Four major groups of
scheduling decision rules are currently contained in the
knowledge base. They are: (1) general requirements; (2)
time; (3) logic; and (4) costs.

First, the general requirements group consists of rules
for verifying activity I-J numbers, activity descriptions
and codes, government activities, local conditions,
participation of major subcontractors, etc,. Second, the
time-related rule group verifies that the project meets
the overall completion date, that the activities durations
are reasonable, that procurement lead time is an integral
part of the construction schedule, that the activity's
float has not been improperly manipulated, etc,. Third,
the group that captures logic-related concepts checks to
insure that submittal activities precede
construction-producing activities, that procurement
activities occur after the proposed materials are
approved, that foundations are completed before the roof
is constructed, and so forth. Finally, the group of
cost-related constraints verifies that the monetary value
assigned to individual activities totals the contract
amount, that the monetary value of each activity
represents a reasonable amount for that work, that the
project has not been excessively front-end loaded, etc,.

Attempts to represent project management knowledge with
rules of an "if ... then" form as well as with the use of
frames has been proven successful by other researchers [5,
6, 7]. The construction knowledge contains heuristics to
assess the effect local conditions may have on the work.
For example, built-up roofing should not be scheduled
during winter if ambient temperatures are expected to be
below specified minimums.

Knowledge to revise the remaining estimated duration of
unfinished activities is also incorporated. For example,
when it is known that several completed or in-progress
activities of the same general work trade or class are
taking longer than originally anticipated, the remaining
duration of all unfinished activities of this class should
be increased as well. This same concept can also be used
for activities that are taking less time [5]. A class in
this context is understood to be an abstract concept that
defines the behavior of its members. Examples of such
classes include but are not limited to: Concrete class,
Formwork class, Electrical class, Subcontractor X class,
Southwest wing class, Third floor class, and so forth.

In addition, the knowledge base contains rules
necessary to assess the reasons for lagging construction.
These rules identify risk occurrences, i.e. set of events
or conditions, that are capable of causing an activity to
deviate from expectations. An in-depth discussion of the

use of risk factors, categories, occurrences, and events for explaining the behavior of a project is found in [7].

Expert System environment

A basic expert system consists of the knowledge base, the inference engine, and the user. The trend, however, is toward more complex, sophisticated, and integrated systems like the one shown in Figure 1. The benefits of being able to access an existing database by far outweigh the complexity and cost of its implementation. In fact, the use of information is now limited more by hardware memory constraints than by its lack of accessibility by other programs.

As shown in Figure No. 1, standard user supplied initial schedule or project progress data are first processed by PrimaveraTM, a highly interactive microcomputer based project management system, to generate five customized reports. These reports contain the minimum information necessary from which inferences will be drawn. For each activity in the project, the following data items are generated: I-J numbers, original duration, remaining duration, percent complete, description, early start, early finish, late start, late finish, total float, free float, code, predecessors, budget, actual dollars spent to date, actual dollars spent this period, estimated dollars to complete, forecast, variance, earned value, etc,.

Once the reports are created, they are automatically loaded onto dBASE-IIITM, a microcomputer based relational database management system. The database environment is designed to host not only the specific project data, but also to accommodate the other user-defined databases with non-project dependent information. Such information includes hourly rates, productivity rates, general economic factors, contractor information, among other things. This feature allows the inference mechanism to analyze the project within a contextual basis.

A statistical module is incorporated to contrast project progress data against the original project plan. These statistical analyses include tests to detect sustained activity slippage, to identify significant differences between actual and planned durations, and to highlight float trends of non-critical activities, among other things. A database within dBASE-III hosts these results.

Personal Consultant PlusTM, a microcomputer based expert system shell, is utilized to emulate the project manager decision-making process. Data from dBASE-III, user supplied project-specific information, and the knowledge base provide a robust setting to this inference

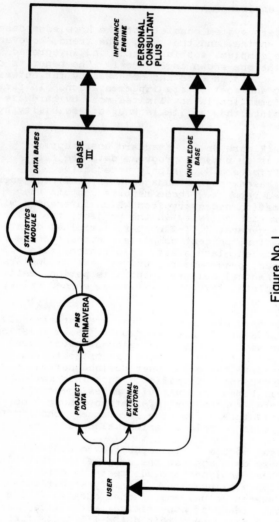

Figure No. I
Expert System Environment

engine. That is. information that is frequently changing can be stored in and accessed from an existing database. This allows the inference engine to retrieve information only on "as needed" basis as opposed to having to embed all these facts in the knowledge base.

Formalism for encoding the knowledge

The core knowledge components of Personal Consultant Plus are:

a) FRAMES to organize knowledge into units/classes:

b) PARAMETERS to represent the facts associated with a given situation:

c) RULES to enforce consistency and specify how additional factors can be determined:

d) PROCEDURES to express algorithmic knowledge processes:

e) ACCESS METHODS to help apply knowledge at the appropriate time: and

f) META-RULES to control the application of other knowledge.

The language employed by Personal Consultant Plus for these descriptions imposes a specific conceptual framework on the problem and its solution [9].

Table No. 1 illustrates a few English-like rules that apply in a variety of contexts. The parameter values are either inferred by backward/forward chaining process. inherited from a higher level class. retrieved from the database. or supplied by the user.

The frame-based representation facility of Personal Consultant Plus is used not only to organize the knowledge base into related units or classes but also to describe the type of classes the system must model. The description of a class (i.e. CONCRETE) represents a prototype description of its individual subclasses (i.e. FOOTINGS. SLABS, COLUMNS, etc.). Classes are defined as a specialization of a more general class. For example, the FOOTINGS class can be described as a subclass of CONCRETE plus additional properties that distinguish footings from other kinds of concrete related tasks (i.e. slabs, columns. and so forth). Our problem domain needs then to be organized as a hierarchical tree of frame descriptors. Each frame descriptor contains a generic specification of

IF   the STATUS of ACTIVITY$_{ij}$ is either <u>in progress</u> or
            <u>finished</u>; and

            its OVERALL ASSESSMENT is either <u>delayed</u> or <u>slow
            progress</u>; and

            its CODE is one of {<u>concrete</u>, <u>formwork</u>,
            <u>electrical</u> .....}; and

            quantity of ACTIVITIES with same CODE, STATUS &
            OVERALL ASSESSMENT > 5

THEN   for all ACTIVITIES with same CODE and STATUS
            other than FINISHED conclude that:

            NEW DURATION = CURRENT DURATION x <u>delay factor</u>

IF   the CODE$_{ij}$ is one of {<u>roofing</u>, <u>concrete</u>, <u>exterior
            closure</u> ...}; and

            LOCATION of the project is <u>mid-west</u>; and

            ACTIVITY$_{ij}$ is WEATHER SENSITIVE; and

            or November < ES/LS$_{ij}$ < March;

                November < EF/LF$_{ij}$ < March;

                ES/LS$_{ij}$ < November and EF/LF$_{ij}$ > March

THEN   Inform the user of possible weather related
            delays

IF   ACTIVITY$_{ij}$ is affected by {<u>snow</u>, <u>cold</u>, <u>heat</u>,
            <u>rain</u>, ...}; and

            SEVERE WEATHER CONDITIONS <u>are present</u>; and

            EVENT DURATION > 0.5 x FREE FLOAT

THEN   conclude that ACTIVITY$_{ij}$ is WEATHER SENSITIVE

Table No. I
Representation of Construction Knowledge

a class in terms of the parameters that are used to represent it and the rules that can be applied to it.

Figure No. 2 illustrates a tree structure for a typical project. In this scenario, the ACTIVITIES frame represents the class of all activities, and the SOUTHWEST FOOTINGS frame represents a specific footing that is a member of the class FOOTINGS.

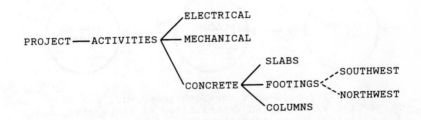

Figure No. 2
Tree Structure of a Typical Project

Status

The performance of the Construction Schedule Analysis system depends on the depth and breath of the knowledge it contains. Although the goal is to develop a product for use by Corps of Engineers field engineers, some parts of the system are still under development.

A subset of the envisioned features has been successfully implemented in both Personal Consultant and Personal Consultant Plus.

Ongoing and Future Research

Complex and ill structured problems such as construction schedule analysis are the domain of Artificial Intelligence. Thus, the development of an effective system is an evolutionary and exploratory process.

It is agreed by the AI community that the power of expert systems lies in the knowledge. Thus, a major effort is being devoted to the expansion and refinement of the current knowledge base.

Future research includes the creation and/or refinement of classes from which activities can inherit self creation behavior. These classes will be defined in terms of the necessary activities that must precede and succeed members of such classes. For example, the addition to the network of an activity such as "placing concrete on column C-9" should automatically generate the following sequence of activities:

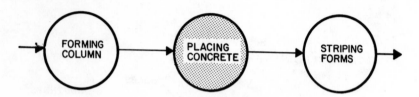

Undoubtedly, this self creation power will not only enhance the verification capabilities of the system, but also will lead us to take our first steps towards automated network generation.

Conclusions

Work to date has shown that modeling construction schedule decision-making requires symbolic processing, the use of heuristics, and decisions based on incomplete and sometimes uncertain information. Furthermore, we have also found that the retrieval and use of frequently changing facts from an existing database opens the door for an integrated system.

Although this work has been on a prototype and thus, can only be seen as a "proof of concept", the demostrated potential suggests that a continued development is justified.

Appendix. - References

1. Feigenbaum, E. A., Davis, R., Buchanan, B. G., and Fox, M. S., "Knowledge-Based Systems and their Applications," _Artificial Intelligence Satellite Symposium_, Texas Instruments, November 1985.

1. Fikes, R., and Kehler, T., "The Role of Frame-Based Representation in Reasoning," _Communications of the ACM_, Vol. 28, No. 9, September 1985, pp. 904-920.

3. Genesereth, M. R., and Ginsberg, M. L., "Logic Programming," _Communications of the ACM_, Vol. 28, No. 9, September 1985, pp. 933-941.

4. Hayes-Roth, F., "Rule-Based Systems," _Communications of the ACM_, Vol. 28, No. 9, September 1985, pp. 921-932.

5. Levitt, R. E., and Kunz, J. C., "Using Knowledge of Construction and Project Management for Automated Scheduling Updating," _Project Management Institute_, Vol. XVI, No. 5, December 1985.

6. McGartland, M. R., and Hendrickson, C. T., "Expert Systems for Construction Project Monitoring," _Journal of Construction Engineering and Management_, ASCE, Vol. 111, No. 3, September 1985, pp. 293-307.

7. Nay, L. B., and Logcher, R. D., "An Expert System Framework For Analyzing Construction Project Risks," _Tech. Rept. R85-4, Order No. CCRE-85-2_, MIT Department of Civil Engineering, Cambridge, Mass., February 1985.

8. O'Connor, M. J., Colwell, G. E., and Reynolds, R. D., "MX Resident Engineer Networking Guide," _Tech. Rept. P-126_, USA CERL, April 1982.

9. Turpin, W., "Personal Consultant Plus," _Tech. Rept._, Texas Instruments, 1985.

Application of Expert Systems to Construction
Management Decision-Making and Risk Analysis

By Roozbeh Kangari,[1] M. ASCE

ABSTRACT: Application of expert systems in the area of construction
management decision-making and risk analysis are explored.
Decision-making under uncertainty is one of the attributes of human
intelligence. Contractors use rules of thumb and subjective
evaluation to analyze a construction project's uncertainty elements.
Traditional construction risk analysis models based on algorithmic
programs have not fully incorporated the significance of empirical
knowledge. As a result of this deficiency, the practical application
of these models have been limited. A successful construction risk
management system requires a significant amount of empirical input
from construction experts and specialists. The main objective of
this paper is to present the potential applications of knowledge
based expert systems in risk management. A microcomputer rule-based
expert system is introduced. Difficulties in developing the system
are discussed.

INTRODUCTION:
     The construction industry is a dynamic field with a high rate of
business failure among contractors. Major causes of the failures
according to Dun and Bradstreet are directly related to management
problems. Construction engineering and management involves many
complex decision-making problems in such areas as resource planning,
cost estimating and control, contractual and legal, political and
public, and other construction management related problems. The
analysis of risk associated with each of these factors is a topic of
great practical interest, because these risks are potentially serious
and have high financial and social impact on major parties involved
in the project.

     Traditionally, construction risk analysis models are developed
based on algorithmic analysis and optimization programs. In these
models the creative component of the construction risk analysis has
been largely ignored. In the real construction world, a large number
of decision-making rules are not based on the mathematical law, but
they are based on the contractor's assumptions, limitations, rules of
thumb, and management style. Contractors use rules of thumb and
subjective evaluations to analyze the uncertainty factors for solving
the problems. This is basically due to the fact that construction
industry has an ill-defined and ill-structured environment. At every

[1]Asst. Prof., Construction Eng. and Management Program, School of
Civil Engineering, Georgia Institute of Technology, Atlanta, GA
30332.

level of decision-making, an engineer has to rely on his judgement and expertise.

The traditional algorithmic risk analysis and evaluation models have not fully incorporated the significance of empirical knowledge. As a result, model users are usually faced with intractable questions. This deficiency has made it impossible for contractors to practically implement these algorithmic models successfully.

A successful construction risk management system requires a significant amount of empirical inputs from construction experts and specialists. This information can be denoted as empirical knowledge, which includes heuristic rules, expert opinions and inferences, and rules of thumb. Empirical knowledge plays a major role at every stage of construction decision-making. Many risk analysis models have failed due to the lack of expert support for novice or semi-experienced model users.

Recent advances in Artificial Intelligence have created new opportunities for solving ill-defined construction management problems by Expert Systems (2, 4, 6, and 7). This allows risk analysis models to combine the classical, physically-based type knowledge, with the much larger body of empirical, rule-based knowledge. This is an important step in the direction of more advanced and practical construction risk analysis and evaluation models.

OBJECTIVES AND SCOPE:
    The objectives of this paper are to explore the implementation of expert systems in construction risk analysis, and to develop a prototype expert system for decision-making under uncertainty. This paper is focused on risk analysis from contractors viewpoint.

During the next decade, the field of expert systems will have an impact on all areas of construction management where knowledge provides the power for solving construction engineering and management problems. The first and most obvious advance will be the development of construction management knowledge base which converts professional construction knowledge into an efficient and productive industrial field. The second benefit is that expert construction systems will catalyze a global effort to collect, codify, exchange, and exploit applicable forms of construction engineering and management knowledge.

In recent years, researchers in the construction field have shown a strong interest in implementing artificial intelligence techniques into the construction field. Expert systems have been implemented in the following construction related areas: pump repair; well selection; structural design; change order evaluation; advice on quality control; estimating the safety practices of contractors seeking bonding; claims analysis; planning and scheduling; construction robotics; etc. The common objective of these expert systems is to absorb technical knowledge from experts, apply it to various situations, and reach conclusions (8,9,11,14,15, and 16).

APPLICATION OF EXPERT SYSTEMS IN CONSTRUCTION RISK ANALYSIS:
Let us look first at the characteristics of construction risk management which makes expert system development possible. One of the most important characteristics is that people with an extremely high level of expertise exist in the construction management area. They have many years of professional construction experience, and can provide the knowledge necessary to build the expert systems. Many of these people are able to articulate and explain the methods that they use to analyze risk.

The second aspect which makes construction risk management appropriate for expert system is that risk management is cognitive, and does not require physical skills. The third characteristic is that risk analysis is not too difficult or too complex for a knowledge engineer to approach. Many sources of information (e.g., papers, books, and etc.) are available in this area which can establish the basic framework for knowledge acquisition.

There are many reasons which justify expert system development effort in risk analysis. First, risks associated with construction projects are potentially serious and have financial and social impact on contractors; therefore, there is a reasonable possibility of high payoff when an expert system is developed. Second, for an inexperienced engineer an expert system can act as an advisor when the professional experts are unavailable. Expert systems are justified especially when significant expertise is being lost to a construction company through personnel changes (e.g., retirement, job transfers, etc.). Finally, expert system development in construction risk management is justified because of the dynamic, ill-structured, and high risk environment of construction field which requires quick decision-making (9,10,13).

In summary, construction risk management is an appropriate area for expert systems since it requires symbolic reasoning, and it is heuristic in nature, that is, it requires the use of rules of thumb to solve problems. Risk management is not easy to model, it takes a human years of study or practice to achieve the status of an expert. Finally, it is of manageable size to be handled adequately by expert systems, and it has a practical value.

COMPONENTS OF RISK MANAGEMENT EXPERT SYSTEM:
Figure 1 shows a schematic view of the construction risk management expert system. The system helps contractors to identify the uncertainty factors, and provides a risk index for the overall project. The system is implemented in INSIGHT 2, a microcomputer knowledge engineering language for rule-based representation. The knowledge base systems created with INSIGHT 2 are complete knowledge and information processing systems capable of applying heuristic knowledge to direct the control and management of conventional programs and data bases. The system's control structure makes use of backward and forward chainings. Numeric data can be expressed as

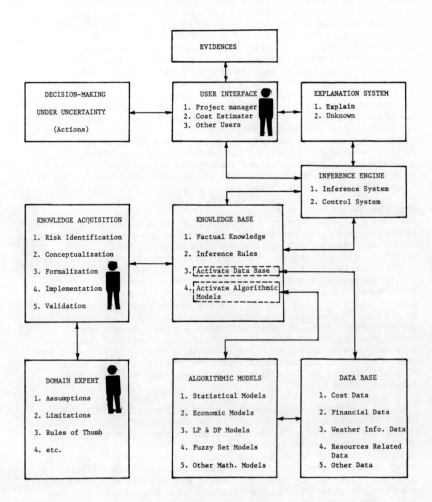

FIG. 1.- Basic Structure of Risk Management Knowledge
Base Expert System

real numbers and variables and can be manipulated with the basic
relational and arithmetic operations. Each supporting condition and
conclusion of a rule can have its own confidence factor. Each
knowledge base has a variable threshold of acceptability which is
used to evaluate the viability of a path of reasoning. The system
also provides the facility for the activation of other programs
written in any language during the execution of a knowledge base.
The system also contains an interface to a Pascal programming
environment which is extended to support direct access to dBASE II
data base files. Through this interface, the system provides a
rule-based expert system the capability to use the power of Pascal
(or other languages) for complex algorithmic computing as well as
relational data base access and manipulation. The system is capable
of handling up to two thousand rules which is considered sufficient
for risk analysis. A portable microcomputer was used during the
interviews with contractors and bonding companies to get feedback
from experts. The major components of the risk management system are
described in the following section.

Risk Management Knowledge Base:
     The knowledge base is a repository of basic knowledge and rules
of construction risk management. The knowledge was collected from
three basic sources: 1) Interviews with contractors; 2) Journal
papers, and 3) Text books. The basic framework was established based
on the last two parts, and the system was modified by contractors.
At the beginning, a small system was developed, and then
incrementally a significant testable system was built. First, the
specification of goals, and constraints was defined. Second, a
general description and classification of construction risk, in terms
of hypotheses, data, and intermediate reasoning concepts was
constructed as shown in Figure 2. Third, the identified elements
were represented in a rule-based (IF-THEN) format (1,3,5,12, and 17).
Then, the system was tested against more complex and real cases.
Many adjustments of the elements and their relationships were a
result of these tests.

     All the contractors interviewed had at least ten years of
construction experience and a yearly volume of less than $50 million.
One of the most important considerations of every firm was the amount
of time allowed by the owner for completion as related to the type
and amount of liquidated damages in the contract. A contract with
heavy liquidated damages combined with a very short time allowed for
completion presented a large and almost unacceptable risk. Another
item was the client/contractor relationship. A good strong client
relationship was highly valued and sought after when the contractor
is considering new work. Such a relationship will mitigate or
dramatically lessen a contractor's risk in such areas as poorly
written contract language or vague project drawings. Other
contractor considerations also included existing workload, repeat
clients, and project location.

     The foremost consideration of the surety companies is the
financial stability of the general contractor. A contractor wishing
to become bonded for a project or to increase his bonding capacity

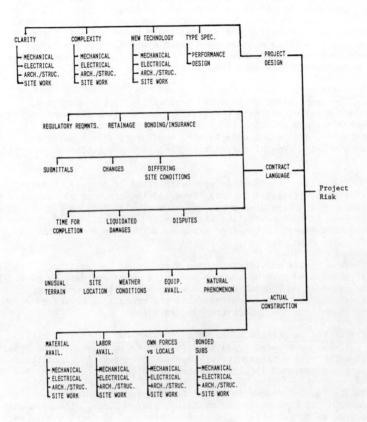

FIG. 2.- Inference Chain of Uncertainty Elements

must furnish the surety with copies of his company's financial
statements, letters or proof of credit, and tax records. Next, the
surety looks at the company's performance history. Such things as
bankruptcy and failure of performance on previous jobs, and size of
jobs performed by the contractor over the last few years. The surety
will then want to know who the general contractor will assign as the
project supervisors for the job and their technical qualifications.
The type of work is also another major element. The results of these
interviews were then presented in rule-based format to enhance the
risk management knowledge base.

User Interface:

The user interface provides capability for the user (e.g.,
project manager, estimator, and others) to monitor the performance of
the system, and provides input concerning construction work
conditions, sources of uncertainty, confidence levels, cost and
economic data, type of contract, information about subcontractors,
and etc.

The system provides the user with trace or display of system
operation by listing all rules fired and the names of all subroutines
called. The system also provides menus of risk factors for the user
to select from when inputing the requested information. The risk
management expert system also explains to users how it reached
particular conclusions.

This system provides the user a mechanism for editing. This is
just a standard text editor for modifying rules and data by hand. It
also provides an UNKNOWN function which allows the user to bypass the
unknown questions, and continue with the process of trying to reach a
conclusion from the information it can obtain.

DIFFICULTIES IN DEVELOPING AN EXPERT SYSTEM FOR RISK MANAGEMENT:

There are several types of difficulties in developing an expert
system for risk management: 1) lack of resources; 2) limitations of
expert systems, and 3) time required to build.

The lack of resources needed for the job consists of two
elements: a) Personnel; and b) Expert system tools. The personnel
required to build an expert system consists of at least a knowledge
engineer and an expert. It was found that knowledge acquisition for
risk management is the most difficult part of the system. The
knowledge engineer must be familiar with the concepts of construction
management, and the expert should provide a sufficient time to build
the system. The expert system tools are still in the stage of
development and only few of the high-level support tools and
languages are fully developed or reliable. In fact, many of them are
new and untested.

Current expert systems and expert system tools have limitations,
many of which will gradually disappear as AI researchers advance the
state of the art. Expert systems have a very narrow domain of
expertise and hence their operation is not as robust as the users

might want. They also have difficulty dealing with inconsistent
knowledge.

Finally, building an expert system takes time. The actual time
required to build a system depends on problem complexity and number
of people assigned to the effort.

SUMMARY AND CONCLUSION:
Construction management risk analysis expert system is developed
to assist management with complex decision-making under uncertainty.
Expert systems will make it possible to develop quick answers for
management problems. It will also help contractors to solve their
productivity problems. Contractors can reorganize themselves into
more efficient and effective organizations. Management will be able
to monitor projects more effectively, and secure their profit by
managing and forecasting the uncertainty factors. In summary, the
construction industry will become much more rational. More
information will be gathered, synthesized, and put into useful form
more rapidly than has ever before been possible.

During the next four to five years many small knowledge systems
and narrow expert systems will be developed for decision-making under
uncertainty. The more complex hybrid systems, the integration of
natural language, and the development of intelligent workstations
that incorporate a large number of different knowledge systems in a
microcomputer will be developed at the end of this decade.

APPENDIX I. - REFERNCES

1. Andriole, S.J., (ed.), "Application in Artificial Intelligence,"
   Petrocelli Book, 1985.

2. Barr, A., and Feigenbaum, E.A. (eds.), "The Handbook of
   Artificial Intelligence," Volume I, William Kaufmann, Inc.,
   1981.

3. Bellman, R., "Artificial Intelligence," Boyd and Fraser, 1978.

4. Bramer, Max, and Dawn, "The Fifth Generation," Addison-Wesley,
   1984.

5. Feigenbaum, E.A., and McCorduck, P., "The Fifth Generation,"
   Addison-Wesley, 1983.

6. Harmon, P., and King, D., "Expert Systems: Artificial Intelligence in Business," John Wiley and Sons, 1985.

7. Hayes-Roth, F., et al., "Building Expert Systems," Addison-
   Wesley, 1983.

8. Kangari, R., "Robotics Feasibility in the Construction
   Industry," Paper presented at the 2nd Conference on the Robotics
   in Construction at Carnegie-Mellon University, June 1985.

9. Kangari, R., "Expert Construction Process Operation Systems
    and Robotics," School of Civil Engineering, Georgia Institute of
    Technology, Technical Report No. 102, March 1985.

10. Kangari, R., and Boyer, L.T., "Project Selection Under Risk,"
    Journal of Construction Division, ASCE, Vol. 107, December,
    1981, pp. 597-608.

11. McGartland, M.R., and Hendrickson, C.T., "Expert Systems for
    Construction Project Monitoring," Journal of Construction
    Engineering and Management, ASCE, Vol. III, No. 3, September
    1985, pp. 293-307.

12. Michie, D., "Introductory Readings in Expert System," Gordon
    and Breach Science Publishers, 1982.

13. Negoita, C.V., "Expert Systems and Fuzzy Systems," The Benjamin
    Cummings Publishing, Inc., 1984.

14. Rehak, D.R., and S.J. Fenves, "Expert Systems in Civil Engineer-
    ing, Construction and Construction Robotics," Annual Research
    Review, Robotic Inst., Carnegie-Mellon University, March 1985.

15. Rehak, D., "Expert Systems in Water Resources Management," Proc.
    ASCE Conference on Emerging Techniques in Storm Water Flood
    Management, Oct., 1983.

16. Sriram, D., "A Bibliography on Knowledge-Based Expert Systems
    in Engineering," CERL, Carnegie-Mellon University, SIGART,
    July 1984.

17. Winston, P.H., "Artificial Intelligence," Addison-Wesley
    Publishing Co., Reading, Mass., 1977.

DESIGN OF AN EXPERT SYSTEM FOR THE RATING OF HIGHWAY BRIDGES

Celal N. Kostem*, M.ASCE

ABSTRACT

A microcomputer-based expert system used to determine the effects of vehicles and overloaded vehicles on simple span bridges with reinforced concrete deck and prestressed concrete I-beams is presented. The system provides an interface to a database and two finite element programs. The reasons which led to the development of the system and the second generation of the system, which is under development, are discussed.

INTRODUCTION

In civil engineering, and especially in structural and bridge engineering, there exists a number of powerful algorithmic programs. The input required from program to program substantially varies; nevertheless, sophisticated programs tend to require a similar sophistication on the part of the user. This tends to discourage use of the programs by non-specialists.

For select areas there exist databases, usually incomplete, that can be used in lieu of the computer programs referred to herein. Also, in almost all areas of civil engineering there exist codes and specifications, far cruder than the results of the above referred programs and the databases, which must be checked, and checked again. When an engineer faces too many options available to conduct a specific mission, the tendency is to take the approach with which the engineer is most familiar. This approach may not necessarily be the most appropriate. For the options with which the engineer is not fully familiar, a mechanism is needed to guide the engineer; thereby permitting the engineer to use approaches with which he/she is less familiar. An expert system can serve the purpose of providing an interface between the engineer and complex computer programs or databases. The purpose of the system described in this paper is such an interface mechanism. It can also be considered as a preprocessor to complex programs, which can be used by relatively inexperienced users.

---

* Professor of Civil Engineering, Fritz Engineering Laboratory, 13, Lehigh University, Bethlehem, PA 18015.

Problem Statement

In the rating of highway bridges engineers can employ a number of approaches. The accuracy and complexity of these approaches differ substantially, from very simple but crude, i.e. literal implementation of AASHTO provisions (Refs. 9 and 10), to very complex but highly accurate, e.g. three-dimensional finite element analysis (Refs. 4, 5, and 8). Regardless of the approach taken, however, it is indirectly implied that the AASHTO approach should also be carried out (Ref. 10). Noticeable deviations from the AASHTO provisions must be "justified," if the AASHTO approach is not used.

In the rating process the following methods, with the indicated advantages and disadvantages, could be considered.

> (1) AASHTO Bridge Rating Provisions (Ref. 10)
>   (Very simple to conduct, but crude.)
> (2) Grillage Analogy (e.g. Ref. 7)
>   (Accurate prediction of beam behavior, questionable information on slab response, requires computerization.)
> (3) "Three-dimensional" linear elastic finite element analysis
>   (Accurate, highly computer dependent (microcomputer or mainframe), complicated input, "response" in one "brief" interactive computer "session" is impractical.)
> (4) "Three-dimensional" nonlinear finite element analysis (Ref.4)
>   (Accurate, essentially mainframe computer dependent, complicated input, can predict the violation of the serviceability limits and damage, and "response" in one interactive computer "session" is most unlikely.)
> (5) Use of a database on previously rated bridges
>   (Refs. 2, 3, and 5)
>   (The data was derived from in-depth analysis and was validated, even the interpolation will give very good results.)

The problem can simply be expressed as the development of an interface program to any one of the five approaches reported above, and to have input expressed in bridge engineering terminology, rather than in terms of finite element method or database management. Furthermore, in view of the extensive availability of microcomputers and their ease of use the system had to be microcomputer-based.

Past experience with practicing engineers' use of finite element discretization clearly indicated that prediscretized geometries must be resident in the system. This will permit the engineer to use any one of the suggested discretizations, without having the need for mesh generation. In terms of input required it was noted that the least amount of information is needed in the case of the use of the database approach, and the amount of information required for other approaches is comparable.

A major limitation was imposed on the system, which in a sense reflects the domain-specificity of the expert systems. The system was built for simple span highway bridges with reinforced concrete deck and prestressed concrete I-beams. There are a number of reasons for this choice. Extensive research on the overloading of prestressed concrete highway bridges resulted in a large database, referred to as "overload directories" (Refs. 2 and 3). Furthermore, a nonlinear finite element program was developed with simplified input and summary output for inelastic analysis of prestressed concrete highway bridges. Computer program BOVAC (Bridge OVerload Analysis-Concrete) contains extensively tested finite element discretizations (Ref. 4).

In the rating process of the bridges the approach taken slightly differs from the routine AASHTO approach (Ref. 10). The reported system will not indicate that the bridge is rated for "x% of HS20-44." The user inputs the vehicle loading and the system checks if any serviceability or strength criteria are violated. If they are not, than the bridge can "legally" carry the specified loading. Otherwise, the system will indicate which of the criteria are violated, thus this load will require special overload permit application. In the case of overload permit applications, the system indicates what type of serviceability criteria may be violated. It is then up to the user to deny or to issue an overload permit.

System Requirements

Past experience indicated that the interaction with the system will be less problematic if the interaction is menu-driven. A brief description of each menu option eliminated the installation of "HELP?" command assistance features. It was assumed, throughout the design of the system, that prior to the use of the system the user would read a couple-page write-up on the use of the system. Extensive manuals were avoided. This was due to another observation: Detailed, self-contained manuals tend to be thick, and the sheer volume of the manuals, regardless of their design, deter their study.

It was further decided that regardless of the user's choice of the approach for the rating of a bridge, the results from two other "approaches" will always be displayed:

(a) AASHTO approach, and
(b) closest interpolation from the database, i.e. the overload directories.

If the results from the database are deemed to be close enough, a message is displayed indicating that the database result should be better than any other approach. However, if the interpolation is a crude one, than the message indicates that the results are given for comparison purposes. If the bridge dimensions and/or the loading pattern are within 5% of one of the case studies in the data base the flagging indicates that a "case with close dimensions" is found. This also applies to the linear and bilinear interpolations, provided that the new problem on-hand can be bracketed within 5% tolerance. All roundings, where necessary, are made conservatively.

Based on the past experience with the finite element simulation of bridge superstructures, "production mode discretizations" can contain an inevitable 5% error, as compared to highly refined analytical models. Thus, within the context of this system a 5% error is considered to be fully acceptable.

In the coding of the system forward-chaining strategy was employed. Initial trial coding of the system, or more specifically the system-manager, employed BASIC, and later Pascal. However, it was later decided to use structured FORTRAN. The use of this language may very well be against the wishes and suggestions of the computer experts; but it is the preferred language of the developer.

Extension of the System

Another observation pertains to the "dependability" of the input information provided by the engineer. This is not due to the ineptitude of the engineer, but due to the wide scatter of the actual data. For example, field observations made in the sixties indicated the dimensions and characteristics of prefabricated components, i.e. prestressed concrete beams, were consistent with the design drawings. However, the actual thickness of the deck slab varied from 6.5 to 11.0 inches, whereas according to the design drawings it was supposed to be 7.5 inches. In view of such a variation the system developer is experimenting with a 10% error being an "acceptable" limit. With this assumption, it would be possible to find the majority of bridge rating problems simulated, either directly or through interpolation, by one or more of the "cases" in the database.

The incorporation of the above concept would permit the elimination of the initial menu of the system, i.e. no option will be given to the user in terms of which approach they wish to employ. If the system can not locate a case in the database for solution of the problem on hand, it will than automatically switch to in-core linear elastic finite element modeler. Another feature that is being tested is that the user will not be able to decide which prediscretized finite element meshes to employ, when needed. The experience with the bridges vs. loading vs. structural response has provided sufficient expertise as to which type of mesh can give the most conservative results.

Without the above changes, the reported system is an expert system which is driven by the demand or experience of the user. Incorporation of the above changes will not give options to the user, i.e. all the expertise will reside in the program. The reason for the deferral of this action to a later date has been due to a very simple reason: To educate the user. By showing two systems operating side-by-side it can be shown to the user that the decisions arrived at by the system are as good as the ones arrived at through deployment of the engineer's immense experience. As was noted in a paper, engineers still have more faith in test results, as compared to computer-based solutions (Ref. 6). In the same vein, engineers tend to have more trust in programs where they instruct what should be done. Thus the reported system can be labeled as a transition mechanism towards full fledged expert systems.

The System Interfaces

The first effort in this study was to transfer 150 single spaced typed pages of overload directories to a database. The first choice for a database system was an old spreadsheet program used for a variety of purposes. This choice was made merely for the sake of convenience and free availability of the package (Ref. 6). Shortly thereafter a three-dimensional spreadsheet program became available. The database was transferred to this package. The three-dimensional, i.e. rows, columns and pages, nature of this spreadsheet program has distinct advantages. Rows and columns were utilized to describe the bridges, and the pages were used to relate the vehicles investigated. However, this attempt was also short-lived. The lack of graphic display options and the insufficient statistical library of the package prevented its expeditious utilization. Thus, the database was transferred to yet another powerful spreadsheet program; again a two-dimensional one.

For the linear elastic finite element analysis the system manager "referred" to the mainframe-based SAPIV program (Ref. 1). Due to the fine tuning of the finite element program and the finite element discretizations, the execution time in the CDC CYBER-family computer has been fast enough to permit the input of the problem and the display of the results in one "acceptably long" session. It is recognized, however, that a microcomputer-based finite element program will enhance the portability of the total system. The system is currently undergoing revisions to "feed" the data to a microcomputer-resident finite element package.

Nonlinear finite element analysis via program BOVAC was not even considered for microcomputer-based operations. At the completion of the session with the microcomputer-based system, the "data" is funneled to CDC CYBER-family computer. The user is informed of this action when the initial menu is displayed. The turnaround time for this option is dependent upon the "off-hours processing" schedule, i.e. after 9:00 AM the next day.

Grillage analogy and AASHTO-type solutions are conducted at microcomputer level.

Finite Element Meshes

Both for linear elastic and for nonlinear finite element analysis the system employs similar discretization of the bridge in transverse direction, i.e. perpendicular to the axis of the bridge. Nodal points are placed on the following longitudinal lines: edges of the deck slab, beam axes, and the lines between the beams. This indicates that one plate bending element each is used for the overhangs, and two plate bending elements are placed between the beams. The discretization in the longitudinal direction employs the following

rule: The bridge is divided into 8 "strips" with the lengths of 0.2L
+ 0.15L + 0.1L + 0.05L + 0.05L + 0.1L + 0.15L + 0.2L (= L). In this
expression   "L" indicates the span length. With this type of modeling
a three-beam bridge will have 48 plate bending elements with membrane
stiffnesses, and 24 Bernoulli-Navier type beam elements.

A second finite element model, which is also the basis for the
grillage analogy discretization, employs the discretization in the
transverse   direction the same as above. In the longitudinal direction
ten "strips" are used, each having lengths of "0.1L." In the above
discretizations no nodal points are placed eccentric to the plate to
be used in the idealization of the beams. The eccentricity of the
beams is built-in to program BOVAC. With this type of finite element
modeling a three-beam bridge will have 60 plate bending elements with
membrane stiffness, and 30 Bernoulli-Navier type beam elements. If
this discretization is used for the same bridge for the grillage
modeling, there will be 94 beam elements.

A third finite element model is similar to the second model referred
to above, but there exist lines of nodes at the centroids of the
beams. The beams are connected to the deck slab via vertical stub
beams. This approach idealizes the interaction between the beams and
the slab via "vierendel truss." If the above referred bridge example
is used again, the following finite elements will be generated: 60
plate bending elements with membrane stiffnesses, 30 beams to simulate
the bridge beams, and 33 stub beams to provide the connectivity
between the beams and the bridge deck slab.

The system recommends the use of the first model with unequal finite
element lengths in the longitudinal direction. It has been found that
this model is quite accurate near the mid-span, where the adverse
action of the loading is expected. It is always recommended that the
loading should be placed in such a manner that maximum flexural
response of the bridge is generated (Ref. 5).

Input Requirements

In the interactive mode the following parameters are needed to access
the database:

        *Problem name. Up to 60-characters long. The first six characters
        are essential in naming the files.
        *Span length (ft.)
        *Weight of an axle load in the heaviest axle group (kips.). Axle
        spacing should not be less than 4-feet.

In the finite element mesh generation or grillage analogy mode the
required additional parameters are:

        *Span length (ft.)
        *Total number of beams
        *Beam types (AASHTO or Pennsylvania Department of Transportation
        standard shapes)

*Out-to-out width of the bridge (ft.)
*Slab thickness (in.)
*Slab thickness, including the integral wearing surface (in.)
*Compressive strength of the deck slab concrete (ksi.)
*Compressive strength of the beam concrete (ksi.)
*Prestressing strand diameter in sixteenth inches (e.g. 7/16 in. strand is inputted as 7)
*Number of prestressing strands in each beam
*Initial prestress on the strands (ksi.)
*Eccentricity of the strand group (in.)
*Age of the bridge in years

*Intensity of the area load (ksi.)
*Longitudinal and transverse distance of the center area load to the prescribed coordinate system (ft.)
*Longitudinal and transverse dimensions of the rectangular area load (ft.)

As can be seen above, the input for the finite element analysis is expressed in terms of bridge engineering. However, it can also be seen that the amount of information required for the finite element option is quite extensive, as compared to the database search.

Observed Shortcomings

For practicing bridge engineers with limited or no exposure to advanced features of bridge engineering, e.g. three dimensional behavior of the bridge, not only the reported system, but the concepts in general tend to be overwhelming. Thus, the critical issue pertains to the education of the profession on new methods and techniques, an issue which has been repeatedly addressed for many decades. Any statement in regard to 5% possible error in the system is considered by some as a major error. However, the research have showed that through the use of the "live load distribution factors," which are the basis for AASHTO type rating (Ref. 10), the error in question may very well be in the 30-50% range (Ref. 5).

The system's response time to the database approach has been encouraging. However, the finite element solution should be conducted in a mainframe environment. This is due to the execution time required for the program in a microcomputer environment. The response time to one option, i.e. database, is measured in seconds, whereas the response time to the finite element approach is predicted in increments of minutes, more like increments of "10 minutes." This, along with the other reasons, is the reason to migrate to the new version, whereby through the adoption of a larger acceptable error, the system will refer to finite element modeling only in rare cases.

It was observed that the last finite element mesh option given in this paper was used with some very wide bridges with 11-15 beams. Such a mesh tends to have a wide bandwidth, or wavefront. Regardless of the "bandwidth optimization" scheme that can be employed, the execution time of such problems tends to be impractically long for microcomputer-based finite element programs. For a given overload vehicle the adverse response to a 15-beam bridge will be similar to that of a 7-beam bridge. Due to the overselling of microcomputer - based finite element programs, it would be desirable to re-educate the user in terms of what can be practically done on a microcomputer vs. what should be desirably channeled to the number crunching mainframe computers.

Conclusions

It is imperative that expert systems be built in the form of preprocessors or postprocessors to provide interface between the bridge engineer and the sophisticated engineering software. The experimental program reported herein has showed the practicality of this approach. Regardless of the personal preference of the author to solve the problems via the finite element method, the response time for a database search is usually faster than the other alternatives. Under the circumstances, it is strongly recommended that attempts should be made to enhance the databases, not only for the type of bridges reported, but for other types as well.

References

1) Bathe, K.-J., Wilson, E. L., and Peterson, F. E., "SAP IV - A Structural Analysis Program For Static and Dynamic Response of Linear Systems," Report No. EERC 73-11, Earthquake Engineering Research Center, University of California, Berkeley, CA, 1973.

2) Kostem, C. N., "Overloading of Highway Bridges - A Parametric Study," Fritz Engineering Laboratory Report No.378B.7, Lehigh University, Bethlehem, PA, 1976.

3) Kostem, C. N., "Further Parametric Studies on the Overloading of Highway Bridges," Fritz Engineering Laboratory Report No.434.2, Lehigh University, Bethlehem, PA, 1980.

4) Kostem, C. N. and Ruhl, G., "User's Manual for Program BOVAC," Fritz Engineering Laboratory Report No.434.1, Lehigh University, Bethlehem, PA, 1982.

5) Kostem, C. N., "Overloading of Prestressed Concrete I-Beam Highway Bridges," Fritz Engineering Laboratory Report No.434.3, Lehigh University, Bethlehem, PA, 1985.

6) Kostem, C. N., "Attributes and Charactersitics of Expert Systems," Expert Systems in Civil Engineering , (C. N. Kostem and M. L. Maher, Eds.), American Society of Civil Engineers, New York, 1986.

7) O'Connor, C., Design of Bridge Superstructures , John Wiley and Sons, New York, 1971.

8) Peterson, W. S., and Kostem, C. N., "The Inelastic Analysis of

Beam-Slab   Highway   Bridge   Superstructures," Fritz   Engineering
Laboratory Report No.378B.5, Lehigh University, Bethlehem, PA, 1975.

9) Standard Specifications for Highway Bridges , 13th Edition,
American Association of State Highway and Transportation Officials,
Washington, D.C., 1983.

10) Manual for Maintenance Inspection of Bridges , American
Association of State Highway and Transportation Officials, Washington,
D.C., 1983.

RC STRUCTURES UNDER SEVERE LOADS -

AN EXPERT SYSTEM APPROACH

by

T. Krauthammer[1], M.ASCE, and S. Kohler[2]

INTRODUCTION AND BACKGROUND

Structural response under the effects of blast and shock loading conditions is of concern to hardened facilities designers who usually employ advanced numerical techniques, such as finite element codes, for the assessment of protective systems. These procedures are expensive and require significant resources for obtaining adequate results, and the engineer would prefer to employ simple numerical tools for fast, yet accurate, structural analyses. Also, if such procedures were available they would be employed for pre-test predictions and for post-test evaluation of model structures. When post-event assessment is concerned the problem could be quite difficult since the experiments are relativelys hort and the researchers need to reconstruct the sequence of events based on the available data, and by inspection of the test speciments. Naturally, different people may interpret such results differently, according the their knowledge of the subject. This fact is a strong reason for developing a systematic approach for the post-event behavior assessment that can provide rational results, and will simulate the judgement provided by an expert.

Structural behavior under blast and shock induced loads can be divided into two principal groups. The first group includes structures under simulated nuclear environments while the second is composed of structures subjected to localized effects such those associated with conventional explosives. The response of reinforced concrete structures under datamation effectgs was treated in the literature, and extensive discussions were presented in previous publications on the subject (7, and 8). Here, a brief summary will be presented as a starting point for developing the expert system.

----
[1]Associate Professor, Dept. of Civil and Mineral Engineering, University of Minnesota, Minneapolis, MN 55455.
[2]Student, Dept of Electrical Engineering, University of Minnesota, Minneapolis, MN 55455.

96

EXPERIMENTAL OBSERVATIONS AND BEHAVIOR

Structures Under Simulated Nuclear Environment

Data from tests on shallow-buried box-type structures, as presented in several publications (6, and 15), can be combined with results from previous studies on this topic (7, 8, and 9) for providing a description of the systems behavior, as follows.

When the simulated airblast is applied to the soil free surface a corresponding shock wave is induced in the soil which travels vertically downward until meeting with the structure. The structural response was controlled either by shear that may result in a direct shear failure at the supports at a very early time, or by flexure where the roof slab may exhibit membrane effects initially in compression and later possibly also in tension. From the observed behavior it becomes quite clear that in the event that direct shear is the mode of failure the structure will not be allowed sufficient time to develop any meaningful flexural response, and therefore one can justify the uncoupling of the direct shear response from a possible flexural response. Furthermore, one may wish to adopt a reliable model for direct shear in order to be able to represent the observed shear failures, and such a model may not have to include the possible effects of moments and rotations at the shear failure region. The failure surface associated with the direct shear failures was essentially vertical, in the direction of the slab depth, while in a typical punching shear failure that surface would have been inclined to the vertical direction. Therefore, one may conclude that the shear mechanism in the direct shear failures is similar in nature to shear transfer across uncracked concrete interfaces, as summarized and discussed in Reference (11). As for the flexural response, any observed failure occurred at a much later time after the roof slab exhibited significant deformations in the membrane mode (12), and the failures were controlled primarily by fracture and pull out of reinforcing bars. Again, it seems reasonable to separate the direct shear response from the flexural behavior.

Structures Under Localized Conventional Effects

The two structures tested under the effects of conventional detonations were geometrically identical to those tested under simulated nuclear effects. There were some differences in the material properties for steel and concrete, as adequately discussed in (5). Overall, there were five shots for structure 3C, and eight shots for structure 3D. Each shot was performed by the detonation of a 21 lb. (9.5 kg) spherical TNT charge at distances between 2.7 ft (0.8 m) to 10 ft (3.0 m). The charges were buried to be positioned at the structure mid-height, and thus, it was expected that the peak pressures would be measured at the wall slab center while the load on the slab would be distributed symmetrically to that point. The main difference between the structures was that in structure 3C the slab thickness was 5.6 in. (142 mm) while it was 13 in. (330 mm) for structure 3D, and the free span wall slab dimensions were 4 x 16 feet (1.2 x 4.9 m). The structural behavior during these tests were described in Ref. (11), and are summarized as follows.

Structure 3C: Tests 1 through 3 were conducted with the charges located geometrically away from the midlength of the structure, and the structure sustained no structural damage due to Test 1 at a range of 8 ft. (2.44 m).

Moderate cracking was observed at the center portion of the wall section after Test 2 at a range of 6 ft (1.83 m), and the cracks radiated longitudinally along the wall. Test 3 was conducted on the opposite side of the structure at a range of 4 ft. (1.22 m), and this test produced extensive damage to the wall, which was displaced inward approximately 10.5 in. (267 mm). Breaching of the wall was believed to be imminent due to the large displacement and the considerable amoutn of concrete spalling, but response of the wall section appeared to be in a flexural mode. The affected wall section was approximately 4 ft. (1.22 m) high by 8 ft. (2.44 m) long. Test 4 was conducted at a range of 2.7 ft. (0.82 m) from the end wall of the box and produced catastrophic damage to the end wall, which was completely blown away and into the interior of the structure. Test 5 was also at a range of 2.7 ft. (0.82 m) but was located between the midlength and the end wall of the structure. This explosion also produced catastrophic damage, and a large portion of the wall was destroyed, and the reinforcing mat was ruptured and drastically displaced. Tests 3 and 5 bracket the charge range to produce breaching of this structure wall and should lie between 2.7 and 4 ft. (0.82 and 1.22 m).

Structure 3D: Structural damage to Structure 3D was not visible until the completion of Test 6, located at a 4 ft. (1.22 m) standoff distance. (Tests 3 through 5 were attenuation tests on the structure end wall). Test 6 produced minor longitudinal cracks on the structure wall. The explotion effects from Test 7 at a range of 2.7 ft. (0.82 m) were sufficient to produce moderate cracking of the wall, roof slab, and floor slab. Vertical cracks on the wall continued as transverse cracks across the roof and floor, and up and down the opposite wall, indicating that the entire structure responded as a large box-beam. When the charge was moved in to a 2 ft. (0.61 m) ragne, the structure was breached. A gaping hole was blown in the side of the wall, and the wall section was displaced inward to the opposite wall. Response of the structure wall during this explosive loading appeared to be in a shear mode. The block of wall blown in was nearly intact, indicating that very little flexural response occurred. The shearing mode of response appeared to be more typical of a direct shear rather than a diagonal shear.

The observed behavior of the test structures, and information provided in the literature indicate that one should anticipate important differences in structural response between the simulated nuclear and localized conventional loads. Localized effects will tend to cause localized damage while the simulated nuclear loads will tend to affect an entire slab or even the whole structure. The arising differences will have to be highlighted by the numerical approach, and careful attention needs to be paid to the corresponding differences in the structural mechanisms which govern the response.

ARTIFICIAL INTELLIGENCE - EXPERT SYSTEMS

The main concept of artificial intelligence (AI) is the modeling process required for progression through a decision tree, by making use of heuristic search, as discussed in (4, and 14). The process can either be done by starting from a set of initial states and progressing toward the goal-state, by consideration of initial conditions and preconditions that allow to make sensible decisions at each intermediate steps. Or the progression to the goal-state can be executed in a backward manner called "backward chaining":

postconditions are set so that the progression from the goal-state back down to the initial states, meets all the conditions of all the intermediate states in a logical way, until the preconditions of the actions necessary to initiate the chain are met. The program can then devise a plan of action to achieve the goal by reversing the sequence of actions found during the search.

The so called "expert systems" are applications of the heuristic search. The emphasis of an expert system is to develop a search procedure through a decision tree by making use of plausible inferences based on knowledge supplied by the user. The possible inferences are stored as rules in the program. These rules are similar to the rules people use to explain their decisions. One feature can be added when the simulation is executed by a computer: a numerical confidence rating or a probability of truth can be assigned to the information processed by the rules, and it is possible to update the rating according to the evolution of the information through the inference network. The confidence rating is the basis of heuristic controls that guide the program search through plausible inferences, and it is applied by the user for assessing the validity of decisions generated by the program. That allows to partially explain why the program reached a particular decision, information that is of course primordial in building confidence in the results.

The range of application of expert systems goes from the derivation problems to the formation problems. A derivation problem is either an interpretation (such is the case for the application of SPERIL-1 of (3) and for PROSPECTOR (2) or a diagnosis (such is the case for MYCIN and SACON (1)) or a monitoring problem. The class of formation problems covers the planning and the design problems (such is the case for HI-RISE (10)). The three major components of an expert system are: the knowledge base that contains the general knowledge on how to solve the particular problems; the context or global data base that describes the state of the dynamic solution, the description being continuously updated; and the inference mechanisms that manipulate the context using the knowledge base.

Some of the tools required to build expert systems are: the general purpose programming languages, the general purpose representation languages, and the expert system building frameworks. The two most common general purpose programming languages used for the development of expert systems are LISP and PROLOG, that permit representations and manipulations of symbols such as "building" or "beam". LISP is the language that is the most common within the United States. The general purpose representation languages are developed specifically for knowledge engineering. Some of those languages are ROSIE, OPS5, SRL, PSRL, HEARSAY-III, and LOOPS. ROSIE is a rule-base language for which the syntax is similar to the English syntax. OPS5 is designed for AI and cognitive psychology. SRL provides specialized constructs for the representation of knowledge. PSRL combines the features of OPS5 and of SRL. HEARSAY-III provides primitives for developing a blackboard data structure accessed and used by knowledge sources. LOOPS is a general tool for knowledge-based program construction.

The expert system building frameworks are inference mechanisms with an empty knowledge base; a knowledge acquisition module is often provided to assist in developing the knowledge base and to structure the data base or context for a particular application domain. Some common frameworks are:

EMYCIN that was developed from MYCIN (1), and KAS that is the knowledge acquisition module of PROSPECTOR (2).

Buchanan and Shortliffe (1) of Stanford University describe the history and the making of MYCIN. MYCIN is an expert system developed for medical diagnostics that saw the creation and the development of many of the now more common techniques of AI and application to rule-based expert systems. They covered topics related to the origin of rule-based systems in AI, the rules, the knowledge base (knowledge engineering, completeness and consistency, and transfer of expertise), the reasoning under uncertainty, and the explaining of the reasoning.

The research undertaken by Buchanan and Shortliffe's group (1) also showed that the domain-specific knowledge can be successfully kept separate from the inference procedures. The result was the creation of EMYCIN from MYCIN. EMYCIN is a consultant expert system that can provide advice on problems within its domain of expertise. An application to the structural engineering field resulted in the SACON expert system. SACON, developed at Stanford University, provides advice to a structural engineer regarding the use of the finite element program MARC: it develops an analysis strategy that can be implemented on the MARC program. Judgemental knowledge about the structural analysis is represented in EMYCIN in the form of rules, combined with the use of certainty factors that allow to draw inferences using uncertain knowledge.

The following two papers present the application of artificial intelligence to an industrial problem. The geological prospection has been modeled on the computer to create a program that serves as a consultant to the user, guiding him/her along the decision process of determination of the geological nature of a particular field.

Duda et al. (2) and Reboh (13) developed the program PROSPECTOR for the U.S. Geological Survey and for the National Science Foundation. The program contains the application of: inference networks, semantic networks, probabilistic reasoning with propagation of probabilities, introduction of certainty factors, fuzzy sets, and a knowledge base that contains the complete description of different geological fields (type, nature, characteristics, components, likelihood, and probabilities). The user answers the questions of the computer, providing the information requested and the confidence he/she has in that information. If the information can not be provided, the program takes note of it by assigning a certainty factor to the data and/or the values assumed to proceed: the program tries to select the piece of unmentioned evidence that will be most effective in confirming or refuting a match. When information is provided by the user the program will update the probabilities of hypotheses regarding that information throughout the network.

Maher and Fenves (10) covered part of the background of expert systems, and described the different types of systems according to their task, and the control strategies followed to perform that task. Their research resulted in the creation of "HI-RISE", a knowledge based expert system developed for the preliminary design of high rise buildings: the user provides to the program the information regarding the space planning of the new building and the program will respond by furnishing the structural information for a preliminary design. The solution provided by HI-RISE is a description of the feasible structural systems, detailed to the level necessry for selection and further modeling and analysis.

With the collaboration of Furuta and Fu, Yao (3) presented a paper on the potential of expert systems in the field of structural engineering. Their conclusions show some of the limitations of expert systems (which can only be as good as the information that the expert is providing), but especially the strength of the expert systems: possible to learn the reasonning process of an expert, possible to work much faster than the expert when the system is effectively working, opening of the modeling and creativity aspects of structural engineering through the use of a computer well beyond the point reached by the analytic capacities that have been so intensively developed in the past.

AN EXPERT SYSTEM

Based on the information presented previously here one can illustrate the composition and operation of an expert system. In general, an expert system consists of the following components, as illustrated in Figure 1:

1. Knowledge base: contains the information used for the creation of a data set that is necessary for the use of traditional civil engineering techniques, and to guide the subsequent analyses that will be performed on the output results.
2. Inference mechanisms: computer simulation of a decision making process, and of a reasoning process required to follow the methodology of the analysis. Also the inference mechanisms will direct the reasoning process required to generate corrections and modifications of the active data base.
3. Context: the context contains the state of the data and of the solution at all the stages of the evaluation process. The initial data set will be transformed continuously into a dynamic set of data. The results of the successive analyses, that make up the context, are organized by the inference mechanisms.

PRESENT APPROACH

It is clear from the previous discussions that the development of an expert system is not a simple task, nevertheless, the effort can be reduced significantly if an existing expert system shell is used and a suitable data base is added to it. Such programs are available commerically, and can be employed without much difficulties. For the present study an existing shell was employed, and the authors had to create the logical flow patterns which simulated the expert, as illustrated in Figure 2.

The expert system involved here serves two functions 1) to analyze the failure mode, and 2) to assess its damage and advise on further loading capabilities.

The architecture involved in this expert system encompases both a tree-structure and databases. The tree structure is mainly incorporated in the beginning of the expert system. Here the user is asked the general condition of the structure, and upon answering these questions the user has subdefined the problem. Take for example the case that the slab is either completely or partially separated from the supports. Since the failure modes for these two

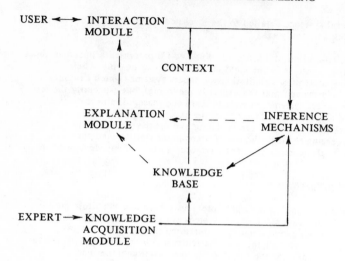

Figure 1. FLOW CHART FOR AN EXPERT SYSTEM (from (10))

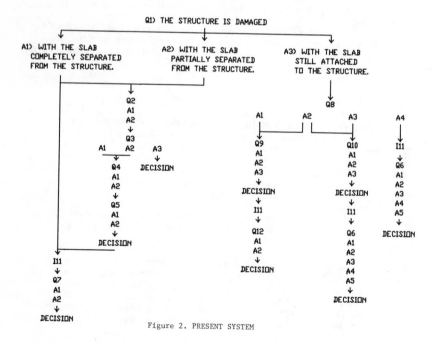

Figure 2. PRESENT SYSTEM

cases are the same, we can now traverse the same branch of the tree.

The user is then asked a series of questions to determine what failure mode operated on the structure. The combination of the succeeding four questions will lead to this decision. However, if the user answers A_3 to Q_3 the system will forego any more questions and will report that the system needs to know a certain aspect of data to obtain a conclusion. Upon reaching the decision of what caused the structure to fail, the system will determine whether or not the failure occurred at all the connections or started at one side first. If the slab was only partially separated the system already knows that the failure occurred at one end first, but if the structure has a completely seperated slab the system continues on to I11 and tests for nonsymmetric failure. After answering an simple question regarding slab position after failure, the system determines if all connections failed at the same time. At this time the system has progressed to the end of the architecture and reports its decisions to the user.

The third option the user could answer states that the slab is still attached at all ends. The user then progresses down to Q_8 when he/she discribes the damage as it performs to the structure. The structure could contain sharp discontinuities and or visible diagonal cracks. Once these symptoms define certain types of failure modes the path is broken up again. Note however that $Q_8 A_2$ leads to two paths, this is of no concern for the system and it does not report any decision until all the rules in the knowledge base have either been eliminated or verified. Questions 9 and 10 require information about the location of the above two symptoms be it in the slab or at the connection regions and from this it deduces the mode of failure. Again the system then enters an intermediate step. Based upon the failure mode (form of shear or membrane action) the system determines the loading capabilities of the structure. This is done through I11, here the system can check for either membrane capacity or shear strength or both, and deduce a decision based upon the following questions. Again after all decisions have been determined through the rules in the knowledge base, the system then reports its findings.

EXAMPLES

Two simple examples are enclosed for illustration of the approach. In both cases the analyst can examine the structure after the test and answer questions, as shown in Appendix II.

CONCLUSIONS

The present approach is an example of the application of expert systems to the post-test assessment of specific structural systems. The results obtained by this approach are consistently accurate and compare well with tohose from other methods. It seems that the present approach is quite promising, and should be developed further.

ACKNOWLEDGMENT

The authors wish to express their thanks to the Defense Nuclear Agency, the Air Force Weapons Laboratory, and the Army Waterways Experiment Station for their cooperation.

APPENDIX I - REFERENCES

1. Buchanan, B. G., Shortliffe, E. H.,"Rule-Based Expert Systems -- The MYCIN Experiments of the Stanford Heuristic Programming Project", Addison-Wesley Publishing Company, October 1984.

2. Duda, R. O., Hart, P. E., Barret, P., Gaschnig, J. G., Konolige, K., Reboh, R., "Development of the Prospector Consultation System for Mineral Exploration", Artificial Intelligence Center, Computer Science and Technology Division, Report prepared for The Office of Resource Analysis, U.S.G.S., and The Mineral Resource Alternatives Program, N.S.F., SRI International, October 1978.

3. Furuta, H., Fu, K.-S., Yao, J. T. P.,"Structural Engineering Application Of Expert Sytems.", School of Civil Engineering, Purdue University, Structural Engineering report CE-STR-85-11. April 1985.

4. Hayes-Roth, F., Waterman, D. A., Lenat, D. B., "Building Expert Systems", The Teknowledge Series in Knowledge Engineering, Addison-Wesley Publishing Company, Inc., 1983.

5. Kiger, S.A., and Albritton, G.E, "Response of Buried Hardened Box Structures to the Effects of Localized Explosions", U.S. Army Engineer Waterways Experiment Station, Technical REport SL-80-1, March 1980 (limited distribution).

6. Kiger, S.A., Getchell, J.V., Slawson, T.R., and Hyde, D.W., "Vulnerability of Shallow-Buried Flat Roof Structures", U.S. Army Engineer Waterways Experiment Station, Technical Report SL-80-7, six parts, September 1980 through September 1984, (Limited distribution).

7. Krauthammer, T., "Shallow Buried RC Box-Type Structures", ASCE Journal of Structural Engineering, Vol. 110, No. 3, March 1984, pp. 637-651.

8. Krauthammer, T., Bazeos, N., and Holmquist, T.J., "Modified SDOF Analysis of RC Box-Type Structures", ASCE Journal of Structural Engineering (to appear).

9. Krauthammer, T., "A Numerical Gauge for Structural Assessment", Proc. 56th Shock and Vibration Symposium, Monterey, California, October 22-24, 1985.

10. Maher, M.L., Fenves, S.J., "HI-RISE: A Knowledge Based Expert System For The Preliminary Design Of High Rise Buildings", Research Report No. R-85-146, Department of Civil Engineering, Carnegie Institute Of Technology, Carnegie-Mellon University, January 1985.

11. Park, R., and Pauly, T., "Reinforced Concrete Structures", Wiley-Inter-science, 1975.

12. Park, R., and Gamble, W.L., "Reinforced Concrete Slabs", Wiley-Inter-

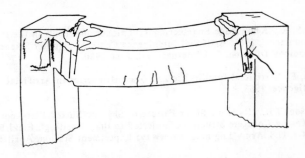

(a) CASE 1

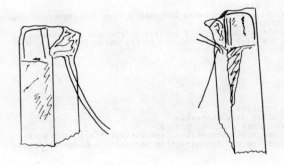

(b) CASE 2

Figure 3. EXAMPLES

science, 1980.

13. Reboh, R., "Knowledge Engineering Techniques and Tools in the Prospector Environment", 1981, Artificial Intelligence Center, Computer Science and Technology Division, SRI International, June 1980.

14. Rich, E., "Artificial Intelligence", McGraw-Hill series in Artificial Intelligence, 1983.

15. Slawson, T.R., "Dynamic Shear Failure of Shallow-Buried Flat-Roofed Reinforced Concrete Structures Subjected to Blast Loading", Final Report SL-84-7, U.S. Army Engineer Waterways Experiment Station, April 1984.

APPENDIX II - EXAMPLES

The following two examples are provided for illustration of the approach. Consider Case 1 of Figure 3(a) in which part of a tested structure is shown. The user interacts with the expert system to reach the final conclusions. Similarly, in Case 2 of Figure 3(b) a different failure mode is detected by the system.

```
                    CASE 1

Enter number(s) of appropriate value(s), WHY for information on the rule
being applied, QUIT to store data and exit or <H> for help

The structure is damaged
     1   with the slab separated completely from the structure and resting on
         the floor.
     2   with the slab separated partially from the structure.
     3   with the slab still attached to the structure.
<REPLY> 3




The structure exhibits
     1   only sharp discontinuities.
     2   only visible diagonal cracks.
     3   both sharp discontinuities and visible diagonal cracks.
     4   neither sharp discontinuities nor visible diagonal cracks.
<REPLY> 3



The discontinuities exist
     1   in the connection regions.
     2   in the slab, away from the connection regions.
     3   both in the connections regions and in the slab.
<REPLY> 1



The visible diagonal cracks exist
     1   in the connection regions.
     2   in the slab, away from the connection regions.
     3   both in the connection regions and in the slab.
<REPLY> 1
```

```
The amount of central deflection observed is
     1    less than one half of the slab thickness.
     2    more than one half, but less than one slab thickness.
     3    more than one, but less than twice the slab thickness.
     4    more than two, but less than half of the slab's free span.
     5    more than half of the slab's free span.
<REPLY> 2




The amount of discontinuity is
     1    less than or equal to .024 inches.
     2    greater than .024 inches.
<REPLY> 2
```

The expert system has determined the mode of failure from the questions you have answered.

Press any key to display results:


```
1    Failure mode in the connection region initiated as Direct Shear.   :1
2    A Flexural response caused the slab to be pushed beyond the peak
     compressive membrane mode and toward the transition to the tensile
     membrane mode.  Future resistance, due to Flexure, should be limited to
     not more than that predicted by the classical yield line theory.   :1
3    Since the Shear Strength is on the declining part of the Shear Stress vs
     Shear Slip curve, reserve capacity could be marginal.   :1
4    The failure mode then shifted from Direct Shear to Diagonal Tension in
     the connection region.   :1
```

          CASE 2

All choices <A>, value>0 <G>, Print <P>, Change and rerun <C>, Quit/store <Q>, rules used <line number>, Help <H>, Done <D>:

```
The structure is damaged
     1    with the slab separated completely from the structure and resting on
          the floor.
     2    with the slab separated partially from the structure.
     3    with the slab still attached to the structure.
<REPLY> 1



The separation has occured
     1    in the slab-wall connection region.
     2    well in the slab.
<REPLY> 1
```

```
The separation between the roof and the structure is
     1    along vertical lines.
     2    not along vertical lines.
     3    unclear.
 <REPLY> 2



The failure surface is defined
     1    by a single surface failure.
     2    by many cracks, "concrete teeth".
 <REPLY> 1



Inspection of the failure region indicates
     1    main reinforcing bars exhibited rupture after significant
          deformation.
     2    concrete was crushed in the compression zone.
 <REPLY> 1



The separated slab, as it is resting on the floor, is
     1    leaning on one side of the structure.
     2    not leaning on one side of the structure, but lying flat on the
          floor.
 <REPLY> 2



1    Catastrophic failure was controlled by Diagonal Tension.   :1
2    This failure occured simultaneously at all the failure regions causing
     the slab to fall to the floor in a horizontal position.   :1
```

DAPS: AN EXPERT SYSTEM FOR DAMAGE ASSESSMENT
OF PROTECTIVE STRUCTURES

T.J. Ross[1] and F.S. Wong[2], Members, ASCE
S.J. Savage and H.C. Sorensen[3]

INTRODUCTION

An automated reasoning code for the damage assessment of protective structures (DAPS) is being developed within the Air Force. Protective structures in this context will be synonymous with buried facilities which are designed to withstand intense impulsive pressures. The code is being written for initial implementation on a microcomputer IBM AT. Two inferencing schemes, both involving back-chaining logic, are being investigated for the DAPS code. DAPS combines crisp numerical data with non-crisp linguistic data using the precepts of fuzzy set theory to estimate the uncertainty in the knowledge base.

The data base comes from a series of eleven experimental tests on buried reinforced-concrete boxes subjected to explosive pressures. Crisp data in the form of instrumentation waveforms and linguistic data obtained from experts through questionnaires dealing with the experimental tests comprise the knowledge base. Figure 1 shows an example of one of the test articles.

DAMAGE ASSESSMENT

The assessment of damage to any structure, civil or military, from any disturbance, natural or man-made, is a difficult process imbued with human judgment. One could justifiably suggest that damage assessment is an art. The process of damage assessment can be broken down into two subsets--damage descriptors and damage level. The former usually involves engineering quantities (although visual images can play a significant role) and the latter involves notions such as function and repairability.

Damage descriptors can be visual but generally are engineering parameters which can be measured. These measured data can be the result of active or passive measurements and can be considered as "hard" data. The visual images referred to earlier can be considered as "soft" data in the sense that they carry much information which has

[1]Senior Research Structural Engineer, Air Force Weapons Laboratory, Kirtland AFB NM 87117-6008
[2]Senior Associate, Weidlinger Associates, Palo Alto CA 94304
[3]Graduate Student and Associate Professor, respectively, Structures Section, Department of Civil and Environmental Engineering, Washington State University, Pullman WA 99164

not been explicitly quantified and identified before. Finally, information residing in the minds of experts who have experience in assessing damage to structures is valuable but inaccessible except through
laborious and painstaking interviews with the experts. This latter
type of information is termed "historical" for this paper. Both the
soft and historical data are typically manifested in the form of expert
opinion and engineering judgment. These concepts are shown schemati-
cally in Figure 2. The procedures used to obtain experts' evaluations
of soft data and to aggregate these opinions in a damage assessment
code are discussed in more detail by Wong (7).

EXPERT REASONING ON DAMAGE

In many fields of engineering, both damage and damage interpretation are not precise. This is especially true of protective structures
because they are heavily reinforced and, yet, are expected to be loaded
to severe damage levels and even to total collapse. Only a limited
number of tests can be performed, and the tests are usually done on
small-scale structures using simulated loading. An example of a
damaged element in a protective structure is shown in Figure 3. The
evaluation of light, medium, and severe damage differs from one expert
to another. Moreover, the damage levels naturally overlap, i.e.,
damage does not change abruptly from light to medium, or from medium to
severe upon reaching certain crisp thresholds. Other factors, such as
scarcity of data and the need to extrapolate the data to realistic
loading, full-size prototypes and imperfect structures, add much more
complexity to the assessment of damage.

Past Experience

For protective structures, past experience in damage assessment
has not always been helpful. Most civil engineers have received
training based on the designs of conventional structures which would
be subjected to natural loads, such as dead loads and wind loads. In
this classic training, damage is usually limited in the design by such
quantities as ductility ratio and structural span deformation. But for
protective structures, some of these engineering damage descriptors
make little sense. Ductility ratio, for example, is a concept which
only has value if the material behaves in an elastic, perfectly-plastic
way, if the elastic yield deformation is known with precision, and if
the boundary conditions (deformations) are also known with precision.
These conditions are generally not known with confidence, especially
for protective structures.

Past experience in determining damage level has also had limitations. The classic case of damage level has been to assign the damaged
structure into one of two groups--safety (survival) or failure. This
assignment process is an arduous task because the demarcation between
survival and failure is not a crisp one, but is rather fuzzy. Even in
the situation where damage is classified into several groupings (such
as, none, light, moderate, severe, and catastrophic), misinterpretation
among experts exists. Some experts assign damage at a certain level
based on the time needed to repair the damage. Others assign damage
level according to the cost of repair, and still others assign a damage
level based on the function of the facility damaged.

Perhaps the most successful (in a qualified sense) damage index employed is the Modified Mercalli Scale used in assessing earthquake damage and, hence, earthquake intensity. But even for this scale, the damage levels are based on experts interpreting linguistically-described damage descriptors (1). There is room for improvement and more consistency in assigning damage levels.

Recent Case Study

A questionnaire concerning damage to the roof element of a buried reinforced-concrete box subjected to a surface blast pressure was mailed recently to 60 experts across the United States. Only pictorial data (photographs) were sent to the experts, and they were asked to provide damage assessments on the structures as well as the reasoning behind their assessments. In particular, they were asked to specify the dominant deformation mode in the roof element and the degree of damage in that mode. This paper will summarize the results of the experts' reasoning for just one structure to illustrate the formation of knowledge rules resulting from the reasoning process.

Figure 1 shows a reinforced-concrete box structure, the roof of which has collapsed completely and rests on the floor. Although this appears to be an easy damage level to assess, results for this structure illustrate the thoughts of experts on the issue of damage.

A majority of experts evaluated the dominant deformation mode in Figure 1 to be shear followed by rebar pullout and loss of anchorage. Other experts mentioned shear without mentioning rebar pullout, which they may have tacitly implied; others mentioned rebar pullout without mentioning shear. These minor differences are probably due to the interpretation of the term "dominant" in the phrase "dominant deformation mode." All experts indicated the degree of damage in the shear mode to be "very severe" or "total" or words similar to these.

The experts used a variety of terms to describe shear. Some examples are: shear, punching shear, shear punching, pure shear, direct shear, vertical shear, edge shear, and sliding shear. With the picture these terms convey the phenomenon equally well, but without the picture the terms may elicit different mental images from different people. For the degree of damage, even more uniformity existed in the experts' assessments (e.g., catastrophic, severe, complete failure, total destruction, collapse). However, without the aid of a picture, these terms still would produce some confusion.

Suppose the assessments on deformation mode and degree of damage are to be incorporated into a knowledge-based computer code. The simple description,

the roof is severely damaged in a punching shear mode with rebars pulled out of their anchor

may not convey the image in the picture depending on the experience of the observer. Equivalently, the predicate sentences,

the deformation mode is punching shear (primary)

```
the deformation mode is rebar pullout (secondary)
the degree of damage is severe (primary mode)
the degree of damage is severe (secondary mode)
```

may be equally cryptic without further clarification. For this purpose, the results on expert reasoning are invaluable. Reasoning will
clarify the assessment and can be used in building a systematic assessment system with computerized expertise. To be able to determine the
impact of reasoning, it is best to first inspect the raw data emerging
from the expert opinions on damage. The following reasons were given
to support the assessment of shear deformation:

```
Failure section is clean
Clean and vertical break with the wall
Slab remains flat, rigid
Sharp deformation gradient near wall
Absence of flexural cracks, large flexural deformations or
    yield lines at center-span.
```

The following reasons were given to support the assessment of
severe/total damage:

```
Structure needs a lot of repair
Structure is not reusable and must be rebuilt
Loss of protective function
Loss of structural integrity
Roof resting on floor
Complete separation of roof slab
Complete damage to (any) interior equipment
No survivors possible
No load capacity left.
```

With this information, the supporting reasoning can be used as
conditions of an expert rule which will lead to the appropriate
inference of the deformation mode and degree of damage. For example,
one rule might read:

```
If slab remains fairly flat
and deformation is concentrated near wall-roof junction
and break with wall is near vertical and clean
and there is an absence of flexural cracks and yield lines
    at center span,
THEN the deformation mode is direct shear.
```

The specific terms used for the deformation mode, i.e., direct
shear, are no longer critical. From the raw expert opinion, a rule has
been created which captures reasoning in the sense that it synthesizes
expertise. More importantly, it provides a better definition of the
deformation mode and eliminates some of the ambiguities inherent in the
terminology.

The same can be done for the assessment concerning the degree of
damage. For example, the raw data provided on damage can be segregated
into three groups depending on which of the following criteria is used:
repairability, functionality, or structural integrity. A rule based on

repairability could be,

> IF the structure needs a lot of repair, or
> the structure is not reusable, or
> the structure must be rebuilt,
> THEN the damage is severe.

A possible rule on functionality is,

> IF loss of protective function is true, or
> any internal equipment is completely destroyed, or
> no survivor is possible,
> THEN damage is catastrophic.

Finally, a rule based on structural integrity could be,

> IF loss of structural integrity is true, or
> the roof is located on the floor, or
> there is complete separation of the roof slab
> THEN damage is catastrophic.

Again, with expertise expressed as rules, the terminology of damage should be evident to the user.

Fuzzy or imprecise terms and phrases, such as "a lot of repair," "fairly flat," and "near vertical," appear inevitable in expert rules. We have exchanged fuzziness at one level (deformation mode, degree of damage) for fuzziness at a lower level (conditions for deformation and damage). Other reasoning conditions, such as loss of structural integrity, absence of flexural cracks, damage to interior equipment, etc., require engineering judgment. To make DAPS more efficient, the fuzziness and judgment are segregated into more fundamental elements.

In general, the results of the recent case study show that experts appear comfortable and positive about giving their opinions, even when the information is incomplete such as that provided in Figure 1. But the information in this figure is sufficient to trigger a firm response from the "experience bank" in the expert. More surprising is the fact that the experts were able to construct a likely chronology of events in the failure of the box, shown in Figure 1, without the active data to support the chronology (obviously a posttest photograph doesn't provide real-time information). A typical example is: initial shear failure, followed by membrane tension until rebar pullout. It is difficult to determine how much of this information is gleaned from the picture or from past experience on these tests (some of the experts knew about the test series). An analysis of the background of the experts, to provide weights to their opinions, is now being accomplished.

DAPS SYSTEM STRUCTURE

The current configuration of DAPS can be described by discussing: what the code is designed to do, what the inferencing mechanism is, how rules are input to the code, and some examples of current rule manipulations. At present, the code uses the process of back-chaining. For example, several attributes of damage are known, such as the information gleaned from Figure 1. The task for DAPS is to back-chain through

the rule-base to match the observed damage attributes with the consequent portions of the rule base to establish which antecedents of the
rule-base are triggered to provide the necessary consequents (a rule is
typically an antecedent-consequent pair, which is also referred to as
an IF - THEN pair).

Requirements

As mentioned earlier, the major inference mechanism is based on a
back-chaining procedure. Another major feature of DAPS is its utility
for approximate reasoning using fuzzy logic. Expert reasoning together
with a rule-based format can circumvent the problems encountered in
using domain-specific terms, such as "shear," and at the same time
synthesize the inherent expertise. An example query in DAPS, such as
the following, would be given when the system encounters such terms as
"direct shear":

> Machine: Does the slab remain relatively flat?
> User: . . . .
> Machine: Is deformation concentrated near the support?
> User: . . . .
> Machine: Is the break between slab and support fairly
> vertical?
> User: . . . .

The code allows other learning, explanatory, and diagnostic tasks to be
performed on the expert rule-base. For example, if the user wants to
know why DAPS comes to a certain conclusion on damage, the code will
provide an audit trail of all the rules which were triggered in making
the decision and whether the data in the rules were part of the code or
were provided by the user during the query session.

From a classification viewpoint, the conditions of the rule for
the example on shear are attributes. When they are matched, the deformation is classified as shear. When they are only approximately
matched, the classification as shear is not definite. The attributes,
matching, and classification involve fuzziness, and fuzzy techniques
are therefore invoked in the DAPS logic.

Finally, an intelligent system should be able to obtain, in lieu
of querying a user, the necessary information about damage to a structure from a photographic medium, such as Figure 1, or from a numeric
medium such as digital waveforms. DAPS is currently focused on the
latter. Plans for future versions of DAPS will involve pattern-and-
vision recognition systems. In a vision recognition mode for the
buried box example, the geometry of the deformed shape of the roof
element plays a dominant role since geometry is the most easily
recognized pattern in the picture in Figure 1. The deformation mode
can be inferred from this pattern. Of course, getting the computer to
recognize this and to use the data is a formidable challenge for the
future. The degree of damage, on the other hand, is more difficult to
assess from visual information alone.

Candidate Expert System Shells

A commercially available shell for producing expert systems is being used in the development of DAPS. This shell is called EXSYS (3). Initial attempts at using an existing expert shell were focused on a damage assessment tool called SPERIL I (4). SPERIL I is a back-chaining expert system written in the C language for a microcomputer. A study of the feasibility of SPERIL I as a shell for DAPS resulted in a version of SPERIL I written in the LISP language for an IBM PC (6). In the present implementation of SPERIL I, the rules as written in the rule-base are formatted such that both the computer and the human may interpret the given commands. A single rule (as will be seen in an example) is actually comprised of several IF - THEN - ELSE type statements. The first line of a rule is the rule number followed by the rule statements, which are written in as many lines as necessary. Each rule is processed in a predetermined order during the inference procedure until a "premise" has been satisfied. At that time, the "action" statement following the "premise" is ignored. An example rule is presented below.

```
Rule0501
        IF:MAT   is    r/c
THEN IF:ISD   <=    -1
   THEN:DRI   uk     1
ELSE IF:ISD   <=    0.4
   THEN:DRI   no     0.9
ELSE IF:ISD   <=    0.8
   THEN:DRI   slig   0.9
ELSE IF:ISD   <=    1.3
   THEN:DRI   mode   0.9
ELSE IF:ISD   <=    2.0
   THEN:DRI   seve   0.9
ELSE IF:ISD   >     2.0
   THEN:DRI   dest   0.8
   ELSE:DRI   uk
```

The variables in the first column after the colons, such as DRI and ISD represent the various damage states and qualifiers used in SPERIL I in the inference process. The expressions in the second column are either mathematical inequalities or degrees of damage (no, slight, moderate, severe, destructive, unknown) which relate the numerical data and fuzzy certainties residing in the third column. Although this type format is concise and efficient in terms of programming techniques, it does not provide the user a presentation of the knowledge base as written in unabbreviated text (although it could be modified to do so). For this reason, another approach has been chosen to represent the knowledge base in the current DAPS code. Rather than spend a great deal of time developing an inference engine and chaining mechanism, it has been found that an expert system "shell" effectively satisfies the present needs.

The "shell" currently being examined for possible use is known as EXSYS. EXSYS is a generalized expert system development package, i.e., an expert system without the knowledge base. EXSYS simplifies the expert system development process, in that it allows the knowledge engineer to concentrate on knowledge acquisition and synthesis.

The rules that can be placed into EXSYS for problem solving are different than those that form the knowledge base for SPERIL I. In SPERIL I, the rules are composed of several IF - THEN - ELSE type statements, while each rule in EXSYS can be formed with a single IF - THEN condition written in natural language or algebraic expressions. The first segment of the example rule given previously is presented here as it would appear in an EXSYS code.

Rule Number: 1

IF: the material of the building is reinforced concrete and (INTERSTORY DRIFT) = -1
THEN: damage due to drifting is unknown with a certainty of 1.0.

TREATMENT OF UNCERTAINTY IN THE KNOWLEDGE BASE

For a rule-based knowledge system, uncertainties may arise from the following sources: validity of rules, applicability of rules, truth of the condition (antecedent), combination of rules and, of course, the manner in which the above uncertainties are treated. This is currently a very active research area, and many approaches have been investigated. A good discussion on this subject is given in Ishizuka, Fu and Yao (5) who have done significant pioneering work in uncertainty analysis in damage assessment.

By contrast, representation of uncertainty regardless of the source is fairly simple in current practice. The uncertainty is usually represented by a single number which has various names, such as confidence factor, certainty factor, weight, probability, and sub-jective probability. Fuzzy sets have also been used to represent uncertainties, and details are given in Wong, et al. (8). Naturally, how uncertainties are represented affects the way they are propagated and combined in a rule-base, and the rule-base formulation itself.

A simple approach to uncertainties is used in the current version of DAPS. Uncertainties can be assigned to the conclusions (THEN parts) of a rule only. They are represented by single numbers in the range (0-1) and are called certainty factors. Several different certainty factors can be associated with the same conclusion. In that event, they are combined using the Dempster-Shafer (2) rule for combining evidences.

Since the standard treatment of uncertainties in EXSYS (the shell on which DAPS is based) is rather restrictive, the following modifica-tions to EXSYS are made. In EXSYS, a probability number is assigned in one of three formats: (0-1), (0-10), (0-100). Uncertainties are combined by taking the arithmetic average, as elements in series ("independent" probabilities) or as elements in parallel ("dependent" probabilities). To adapt EXSYS to DAPS, uncertainty factors are repre-sented by the mathematical variable option in EXSYS in place of the standard treatment. For example, unknown damage is one of the several possible damage states, and is represented by the variable (UNKNOWN). Its value is initially set to zero. When rules related to this damage

state, such as Rule Number 1, are triggered and the damage state is
given an uncertainty factor, this factor is imbedded as the value for
(UNKNOWN). Assignment of another certainty factor to the same variable
(as a result of the triggering of more rules related to this damage
state) will encounter a non-zero resident value for (UNKNOWN). This
will trigger the Dempster-Shafer mathematical operations which are
coded as part of the THEN portion of the rule. (UNKNOWN) is updated in
this manner and as many times as necessary to combine all uncertainty
factors related to a particular damage mode.

SUMMARY

This paper describes the development of an expert system to
accomplish the damage assessment of protective structures which are
subjected to extremely high impulsive pressures. The damage assessment
process involves the use of damage descriptors and defined damage
levels which are highly subjective and variable among experts reporting
on the damage. Expert reasoning on damage, given pictorial data, can
be consistent even for experts from diverse backgrounds. Without this
pictorial data, verbal descriptions of damage can be misinterpreted.
The fact that the assessments on damage reside in the domain of experts
and the notion that subtle differences in data can produce misleading
conclusions points to the need for an expert system which is capable of
reasoning about damage.

DAPS is a back-chaining system using approximate reasoning with
the precepts of fuzzy logic. Most expert systems can only accommodate
exact matching in the triggering of rules in the knowledge base.
Accounting for the fuzzy and imprecise terms which inevitably reside in
expert rules can be done with approximate matching. Finally, DAPS uses
an expert shell to manipulate the rule base and to form new rules from
natural language inputs.

ACKNOWLEDGMENTS

The authors are grateful to the Air Force Weapons Laboratory and
the Air Force Office of Scientific Research for their continuing support of this work.

REFERENCES

1. Boissonnade, A.C., "Earthquake Damage and Insurance Risk,"
    Ph.D. Dissertation, Stanford University, Stanford, California,
    June 1984.

2. Buchanan, B.G. and Shortliffe, E.H., Rule-Based Expert
    Systems, Addison-Wesley, 1984.

3. Huntington, D., EXSYS: Expert System Development Package,
    Albuquerque, New Mexico, 1985.

4. Ishizuka, M., Fu, K.S. and Yao, J.T.P., "SPERIL I Computer
    Based Structural Damage Assessment System," Purdue University,
    Structural Engineering Technical Report No. 81-36.

5. Ishizuka, M., Fu, K.S., and Yao, J.T.P., "Inexact Inference for Rule-Based Damage Assessment of Existing Structures," 7th Int. Joint Conf. Artificial Intelligence (IJCAI), Vancouver, August 1981.

6. Savage, S., "SPERIL.LSP: A LISP Version of SPERIL I, An Expert System for Damage Assessment to Buildings," AFOSR Summar Graduate Student Report, August 1985.

7. Wong, F.S., "Modeling and Analysis of Uncertainties in Survivability/Vulnerability Assessment," Air Force Weapons Laboratory Report, AFWL-TR-85-84, September 1985.

8. Wong, F.S., Dong, W., Boissonnade, A., and Ross, T.J., "Expert Opinions and Expert Systems," Proceedings of the Ninth ASCE Conference on Electronic Computation, Birmingham, Alabama, February 1986.

Figure 1. Sample test article showing damage observed after experiment

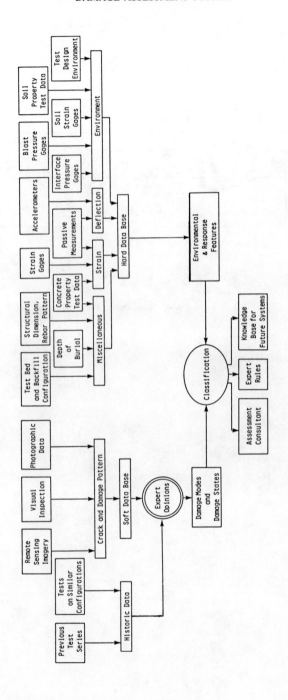

Figure 2. Examples of Historical, Soft, and Hard Data

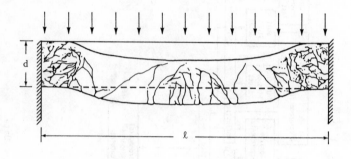

Figure 3. Example of a damaged roof element

Seismic Risk Analysis System

Glenn H. Miyasato,[1]
Weimin M. Dong,[2]
Raymond E. Levitt,[3] *M. ASCE*
Auguste C. Boissonnade,[4] *M. ASCE*
Haresh C. Shah[5] *M. ASCE*

Seismic risk involves many factors such as ground motion, vulnerability of structures and social impact. Most of these are not well defined and are not easy to quantify. However, experienced engineers and researchers do have some judgemental and heuristic rules to combine the factors. The Seismic Risk Analysis System was developed to model this kind of expertise and to provide consultation to the users.

The Seismic Risk Analysis System uses *Deciding Factor*^{TM}, a tool for building expert systems which has a sensible interface with the users. An overall inferential tree was generated to describe the interrelations between all factors. The top node is the hypothesis of total risk. By a backward chaining search, the sub-hypotheses were verified or rejected. The combination rules were composed of most-logic, best-logic, worst-logic, and all-logic to facilitate the fitting of the system to the actual knowledge base. The answer to the inquiry of the users by the system are very flexible. The can be binary (yes or no), discrete (different levels) or continuous values with or without a certainty factor attached to them. The system also provides an explanation regarding how a conclusion was reached.

Introduction

The Seismic Risk Evaluation System is an outgrowth of an ongoing project at Stanford University involving risk analysis and seismic safety of existing structures. A preliminary step of the project was to select from a group of buildings the ones that were suspect in terms of seismic risk criteria. The development of an expert system was one means to accomplish this particular task. Experts at the John A. Blume Earthquake Engineering Center at Stanford University involved in the project were consulted for the knowledge base development.

The intent of this study was neither to develop expert system software nor to advance the state of the art in seismic risk evaluation but instead to use a commercial, readily available expert system development package or "shell".

---

[1] Graduate Student, Dept. of Civil Engineering, John A. Blume Earthquake Engineering Center, Stanford University, Stanford, CA 94305

[2] Graduate Student, Dept. of Civil Engineering, John A. Blume Earthquake Engineering Center, Stanford University

[3] Associate Professor, Dept. of Civil Engineering, Stanford University

[4] Post-Doctoral Research Scholar, Dept. of Civil Engineering, John A. Blume Earthquake Engineering Center, Stanford University

[5] Professor, Dept. of Civil Engineering, Stanford University

The expert system tool *DecidingFactor*$^{(TM)}$ was investigated for use in building a seismic risk evaluation system. The resulting system was then internally validated in several case studies taken from the city of Palo Alto, California.

Seismic Risk Evaluation

The problem of assessing a level of seismic risk of a building is appropriate for expert system development because much of such an assessment involves expert opinion and knowledge from past experience. Also, the fact that many criteria involved in a seismic risk evaluation are not well defined or that measurements are difficult to verify make judgemental and heuristic rules even more important.

Specifically, this study details the use of an the expert system approach to assign a relative risk level to an individual building based on certain seismic risk criteria. Seismic risk is defined as the likelihood of loss due to earthquakes and involves four basic components: hazards, exposure, vulnerability and location. These factors are further defined below.

1.   The hazards or dangerous situations may be classified as follows:

    1.1.   Primary hazards (fault break, ground vibration)

    1.2.   Secondary hazards which are potentially dangerous situations triggered by the primary hazards. For example, a fault break can cause a tsunami or ground shaking can result in foundation settlement, foundation failure, liquefaction, landslides, etc.;

    1.3.   Tertiary hazards produced by flooding by dam break, fire following an earthquake and the like.

    All these hazards lead to damage and losses. They may be expressed in terms of severity, frequency and location.

2.   The exposure is defined as the value of the structures and contents, business interuption, lives, etc.

3.   The vulnerability is defined as the sensitivity of the exposure to the hazard(s) and the location relative to the hazard(s).

4.   The location is defined as the position of the exposure relative to the hazard.

Losses resulting from seismic hazard are numerous and can be categorized as life and injury, property, business interruption, lost opportunities, contents tax base, and other losses. A seismic risk analysis requires the identification of the losses to be studied as well as the identification of the hazards, exposures and their locations and vulnerability. For the purpose of this study, life and injury losses resulting from seismic hazard were the major considerations in the evaluation of the risk level. Some aspects of the four components affecting risk were not included in the evaluation. The parts of the four components that were used in this evaluation are organized as shown in figure (1). The hierarchy starts with the main idea at the top and progresses to the supporting levels below. The main idea appears as: "Seismic Risk Level". At the next level, three key ideas, "Seismic Hazard", "Building Vulnerability", and "Building Importance" support the main idea. These ideas have additional sub-levels of increasingly specific support.

The "Seismic Hazard" idea includes some aspects of the hazards and the location components. The supporting ideas considered below this idea consisted of the primary and secondary hazards. The primary hazard of ground vibration along with the location component was included in the "Ground Shaking" supporting idea. A severity parameter (peak ground acceleration, modified mercalli intensity) was used to measure the ground shaking level. This parameter implicitly included the location of the exposure relative to the hazard. Two secondary hazards, liquefaction and landslides, were selected for inclusion below the "Ground Failure Potential" supporting idea along with the primary hazard of fault break. Due to the fact that landslides tend to occur where the natural grade is relatively steep and where top soil layers are underlain by differing materials or lubricating layers, landslides were further supported by grade steepness and soil geology. The last supporting idea involved was "Soil Type" where extreme soil types can affect the seismic hazard level.

The "Building Vulnerability" idea reflects the vulnerability component by taking into account the sensitivity of a particular structure to the seismic hazard. Thus, supporting ideas below "Building Vulnerability" include structure characteristics such as building type (structural system and type of material used), structural alteration or weakening, quality of construction and seismic design quality. Seismic design quality refers to specific areas in the design which may affect the building's seismic performance. This category includes seismic code considerations, vertical stiffness discontinuities, structural system redundancy and architectural configuration factors such as plan and vertical symmetry, significant re-entrant corners, etc.

The "Building Importance" idea reflects one aspect of the exposure component; the value of human life. "Building Importance" reflects the utility value that is placed on the structure; where utility is measured in terms of public safety. High importance suggests that damage or destruction of the building due to an earthquake would be detrimental to public safety. Therefore the supporting ideas are concerned mainly with how building use and occupancy affect the possible loss of human life during and after an earthquake.

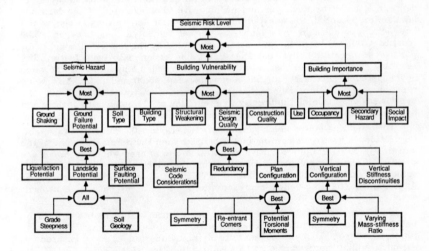

Figure 1. Simplified inference network of the Seismic Risk Evaluation System

Hardware and Software Tools Used in the Seismic Risk Evaluation System

A computer-based damage assessment system has been developed at Purdue University. (Ogawa et al, 1984, Ishizuka et al, 1981). The system uses observation data and other deep knowledge to assess a level of damage to an existing structure that has been subjected to earthquake excitation. This particular study intends to develop a seismic risk evaluation system which assigns a level of risk to a building before an earthquake occurs. Also, much of the knowledge used in this system was heuristic.

The Deciding Factor program was chosen for application since the risk evaluation methodology fitted the framework. The program uses a decision model formulation, where ideas are organized in a hierarchical tree from the general to the specific.

This problem decomposition approach to building an expert system was first employed by the team at SRI International that developed the PROSPECTOR mineral resource evaluation expert system. (Campbell et al, 1982; Duda et al, 1979). Deciding Factor, authored by Dr. Alan Campbell, one of the scientists on the SRI team, is a direct outgrowth of this research.

Deciding Factor is a production rule based expert system shell which contains a backward-chaining inference engine to flexibly connect facts provided by the expert. It is composed of two parts called the $Editor^{(TM)}$ and the $Consultant^{(TM)}$. The Editor guides the builder in the construction of a graphic model of ideas that support a proposed goal. The Consultant directs questions to the user and issues a report on the proposed goal based on the responses to the questions. Because the questions are derived from the ideas, each idea must be a statement that can be either supported or denied during consultation.

All ideas at the bottom of the hierarchy are examined by the Consultant. These ideas, called factors, are converted into questions which are answered in degrees of belief between yes and no inclusive. For example, if the factor is liquifaction potential, the question becomes: to what degree do you believe the liquifaction potential is very high? Numerical values ranging between +5.00 (yes) to -5.00 (no) to each response are multiplied by the numerical weights for positive or negative belief that the expert has assigned to each factor. The resulting product is then posted to the above idea called a hypothesis. The value is truncated if necessary so that it lies between +5.00 and -5.00. This value represents the numerical degree of belief in the hypothesis. Thus, the hypotheses are examined indirectly by evaluation of the supporting factors and hypotheses below them until the top hypothesis or main idea is examined.

The weighting factor which ranges between -1.0 to 1.0 is actually two weighting factors. A positive weighting factor is used when the numerical value of the question is positive and a negative weighting factor is used when the values are negative. This is a significant departure from the Bayesian likelihood factors for sufficiency and necessity used in PROSPECTOR. In this application we found that it simplified the knowledge representation and gave essentially equivalent results.

Each question was also assigned an importance value which was used by Deciding Factor to calculate a reliability estimate of the conclusion. The importance varied from 0 (low importance) to 99 (high importance) and measured the extent to which the user needed to answer a question with a great deal of certainty.

For certain questions, user responses (+5.00 to -5.00) were assigned as categories (e.g. type of structure) instead of degrees of belief. The system prompts the user for a value and refers the user to a table. The user then picks the value assigned to the appropriate category, where the assignments were made by an expert. The assignment of values to these questions replaces the concept of "degree of belief" in the answers. Therefore the reliability estimate as calculated in the conclusion of Deciding Factor no longer gave an accurate measure of the reliability of the final answer. For instance, an answer of 0.0 was taken by Deciding Factor to be a very uncertain response when in fact it corresponded to an exact category (e.g. medium soil). To remedy this, the importance of these "value" class questions was set to zero so that they would not affect the reliability estimate.

The system can be run on an $IBMPC,XT,AT^{(TM)}$, or equivalent computer with at least 128K of memory. Deciding Factor is very easy to use as no external editor is needed to input the knowledge base. Built-in logic screens display relationships between rules to show why a particular question is being asked. Deciding Factor also prints out a decision tree showing the overall logic as well as the weights assigned to each factor. This capability makes it very convenient for knowledge base development. The system also provides the ability to attach text or extended character set graphical explanations to variables or questions. Thus, questions or hypotheses can be clarified or expanded using explanation screens.

Knowledge Representation

Each group of factors below a hypothesis must be linked together so that they can be evaluated and a conclusion can be reached about the hypothesis. Deciding Factor provides several relationships to express the logic between factors. The most appropriate relationship was chosen and fit to each situation.

One such situation shown in figure (2) involved the factors of grade steepness (x_1) and soil geology (x_2) below the landslide hypothesis (y), where both factors support the hypothesis equally. Of the logical relationships provided by Deciding Factor, the 'ALL' relationship was found to be the most appropriate.

For the 'ALL' logic, the sum of the numerical values of the factors $(x_i, -5.00 < x_i < +5.00)$ multiplied by the corresponding weights $(w_i, -1.0 < w_i < +1.0)$ becomes the numerical value of the hypothesis:

$$\text{Equation 1:} \quad y = w_1(x_1) + w_2(x_2)$$

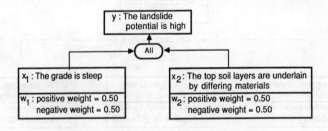

Figure 2. Landslide Hypothesis and Supporting Ideas

The relative weights are normalized so their sum is 1.0. Therefore the strongest support the hypothesis can receive is 5.00 (definite yes) to -5.00 (definite no). Since both factors counted equally towards supporting the hypothesis, w_1 and w_2 were set to 0.5. As an example of this logic, consider the following case. The grade is steep; therefore $x_1 = 5.00$. The top soil layers are underlain by differing materials; therefore $x_2 = 5.00$. The value of (y) is:

$$y = 0.5(5.00) + 0.5(5.00) = 5.00 \quad \text{(the landslide potential is high)}$$

Unequal weights can be used if one factor affects the hypothesis more than the other. Similar differing positive and negative weights account for prior probabilities above or below 0.5 in a Bayesian sense.

Another situation involved evaluating the seismic hazard (y). The expert considered three supporting factors: ground shaking (x_1), soil type (x_2), and ground failure potential (x_3). Figure (3) shows this particular case. The problem involved finding a way to link these three factors which all supported the seismic hazard to some degree. Through the expert, it was found that when the ground motion level is very high, the seismic hazard level would certainly be high regardless of the other two factors. However, when the ground motion level is moderate or low, then the other two factors do contribute to the determination of the seismic hazard level. Therefore, a relationship was sought which weighted some factors more than others and allowed one factor to govern the situation in extreme cases. In order to fit this knowledge structure, the following combination logic was chosen.

$$\text{Equation 2:} \quad y = w_1(x_1) + w_2(x_2) + w_3(x_3)$$

This logical relationship is called 'MOST' in Deciding Factor. The weights are not normalized; therefore their sum can be greater or less than 1.0. However, the numerical answer for the hypothesis is truncated so that it still ranges between 5.00 and -5.00. Since information about ground shaking should govern the evaluation of the hypothesis, w_1 was set to 1.0. The weight w_2 was set to 0.5 and the positive weight for w_3 was set to 0.3. The unequal weights were used because some factors affected the hypothesis more. Factors which could only add support but not reduce support for the hypothesis were given negative weights of 0.0. For instance, since a high ground failure potential can increase seismic hazard but a low potential for ground failure will not decrease the hazard, ground failure potential was assigned a negative weight of 0.0.

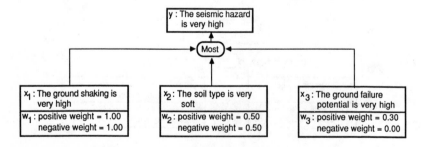

Figure 3. Seismic Hazard Hypothesis and Supporting Ideas

Consider an example of the 'MOST' logic for the following specific case of seismic hazard. The ground shaking level is very high; therefore $x_1 = 5.00$. The soil type is moderately soft; therefore $x_2 = 2.50$. The ground failure potential is very low; therefore $x_3 = -5.00$. The value of (y) is:

$$y = 1.0(5.00) + 0.5(2.50) + 0.0(-5.00)$$

$$y = 6.25 > 5.00$$

$$y = 5.00 \quad \text{(the seismic hazard is very high)}$$

The logical relationship 'MOST' and unequal weights were also used to combine the three supporting ideas below the main idea. The seismic hazard factor was given the highest weight of 1.0 so that it would govern the assignment of a risk level over the other two factors. This weighting gave good results except in the case where hazard was very low and both of the other factors were very high. However the possibility that both of the other factors will be high at the same time is extremely unlikely so for practical purposes the weighting used was appropriate.

The logical relationship 'MOST' was used whenever it was decided that one factor should influence the decision much more than the others. The possibility of using conditional logic (where out of range answers cause pruning or branching in the tree) should be investigated as an alternative to the 'MOST' relationship in some cases. For the remaining cases a relationship was needed where any one factor by itself could prove the hypothesis. For example, ground failure (y) is caused by the occurrence of any one of the three factors: liquefaction (x_1), fault break (x_2), and landslide (x_3). In other words, the most strongly supported factor below the ground failure potential will support it. Figure (4) shows this case.

The relationship used by Deciding Factor that was deemed most appropriate was the relationship 'BEST'. 'BEST' uses the following combination logic:

Equation 3a: $y = \max[(x_1), (x_2), (x_3)]$

In the case where certain factors may qualify as being more important than others, weighting factors can be introduced:

Equation 3b: $y = \max[w_1(x_1), w_2(x_2), w_3(x_3)]$

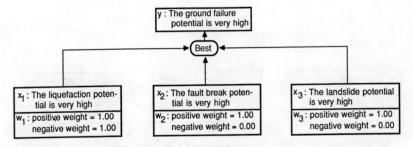

Figure 4. Ground Failure Potential Hypothesis and Supporting Ideas

The following calculation shows the use of the 'BEST" logic with a specific case of evaluating the ground failure potential. The liquefaction potential is very high; therefore $x_1 = 5.00$. The fault break potential is very low; therefore $x_2 = -5.00$. The landslide potential is high; therefore $x_3 = 2.50$. The value of (y) is:

$$y = \max[\,1.0(5.00),\ 0.0(-5.00),\ 1.0(2.50)\,]$$

$$y = \max[\,5.00,\ 0.00,\ 2.50\,]$$

$$y = 5.00 \quad \text{(the ground failure potential is very high)}$$

For most 'BEST' cases, all factors were equally weighted. However, for the building importance case as shown in figure (5), the factors were unequally weighted because each factor was actually a category of building types. The categories with the more important buildings were given more weight relative to the other categories. For instance, essential facilities (hospitals, fire stations, other buildings neccessary for post-earthquake recovery) must function during and immediately after an earthquake, so they are considered more important than secondary hazards (buildings which pose an immediate health hazard if damaged). Therefore, the essential facilities weight is greater than the secondary hazard weight. The fuzzy logic type of flexibility in Deciding Factor was found to be advantageous in this application.

Although the main idea was to determine if the building was a high risk building, the conclusion was not to determine clearly whether or not the building was at high risk, but rather to determine the level of risk. Therefore a grading system based on the final numerical conclusion was developed. For instance, a range of 5.00 to 3.75 indicated a very high risk, a range of 3.75 to 1.25 indicated a high risk, 1.25 to-1.25 indicated moderate risk, -1.25 to -3.75 indicated low risk and -3.75 to -5.00 indicated very low risk.

System Validation

Several structures were evaluated by the system and compared to evaluations made by an expert whose knowledge was used in constructing the system. This internal validation was performed on five buildings in the city of Palo Alto, California. For the purpose of this study, the buildings were labeled alphabetically from A through E.

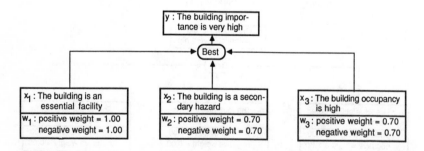

Figure 5. Building Importance Hypothesis and Supporting Ideas

Building A is a nine story office building with approximately 500 occupants during commercial hours. The structure, built in 1972, has a reinforced concrete core wall with cast-in-place interior columns, precast exterior columns, precast beams, and precast floor planks. The building exhibits both plan symmetry and elevation regularity.

Building B is a one story hospital housing approximately 100 occupants during the day and 60 during the night. The structure was built in 1966. It is of mixed construction comprising of concrete buttresses, interior box steel tube columns and wood roof framing. The building is square in plan with buttresses on three sides.

Building C was originally a single story unreinforced masonry building constructed in the 1890's. The structure is used to house a retail outlet and has a daytime occupancy of approximately fifteen people. The building is rectangular in plan and originally had a tin roof supported by timber trusses. However the building was modified by adding a new floor at the truss lower cord level. Timber columns were added to take care of additional loads but show no sign of anchorage to roof trusses or to the existing floor.

Building D, constructed in 1963, consists of a combined shear wall, reinforced concrete column, and wood truss roof system. The building is a fire station with an occupancy of three. The one story structure is composed of a hanger housing the fire trucks flanked by an adjoining room on either side. The building is crossed-shaped in plan.

Building E is a large church constructed in 1959. It holds approximately 150 people on sunday mornings and is usually empty otherwise. The structure is primarily a wood frame with diagonal columns that join at the roof peak. The church is rectangular in plan with a large dormer on one side.

The results of the separate evaluation by the system and by the expert are shown in table (1) and figure (6). The evaluations compared favorably with each other. Most buildings were assigned the same risk category by both system and expert, and both evaluations ranked the buildings in the same order from the highest risk level to lowest.

Table 1

Comparison of Risk Levels			
Building	Internal Expert Assigned Level	System Assigned Level (I)	System Assigned Level (II)
A	low-moderate (-3.75 to 1.25)	moderate (1.1)	moderate (-0.4)
B	moderate (-1.25 to 1.25)	high (1.6)	moderate (0.1)
C	high (1.25 to 3.75)	high (3.2)	high (1.7)
D	high (1.25 to 3.75)	high (3.3)	high (1.8)
E	low (-3.75 to -1.25)	low (-1.3)	low (-2.8)

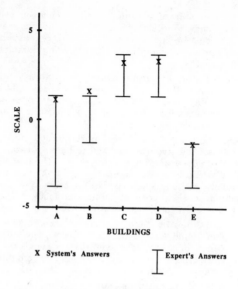

Figure 6. Risk Levels Assigned by Expert and System

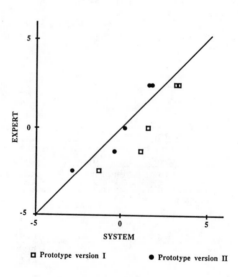

Figure 7. Seismic Risk Evaluation System

The comparison between the answers of the expert and the system represented in table (1) and figure (6) are represented in a scattergram (figure 7). Each interval representing the expert's answer was represented by its mean value. From figure (7) we can see that version I of the system consistently assigned slightly higher risk levels. Calibrations on the weighting factors were then performed such that all points were as close as possible to the straight line passing through the origin point. This resulted in version II of the system.

Extensive validation using buildings of different type and age are currently underway, and this system is currently being considered as an information tool to building owners in the city of Palo Alto, California.

Conclusion

The Seismic Risk Evaluation System provides a preliminary evaluation of seismic risk for different types of existing buildings. The possible applications of this type of evaluation include assessments for possible seismic upgrading of buildings, city planning, disaster mitigation, and insurance feasibility studies. The system could also be implemented as an informational aid to educate the general public about seismic risk. As the system runs on a micro-computer, it is easily accessible and economical to develop and use. The Deciding Factor Shell provides a flexible, easy to use development and consultation package. Besides the consultation, the user can also receive some insight on how the risk assessment is reached.

However, this risk model is by no means complete. First of all, only a limited number of factors affecting risk were taken into consideration by the model. Some factors, such as liquefaction potential, fault break potential, ground shaking level, and quality of construction, could have been expanded several more levels downward. These factors themselves are of sufficient complexity that individual systems can be developed for each of them. Secondly, factors such as building type and seismic considerations, which have assigned values as input need to be improved so that values exist for every possibility. The current values are not compatible with the broad range of observed facts. Some inputs are quite subjective as they are in the form of linguistic descriptions. Fuzzy set theory is investigated as a means to deal with this kind of linguistic information. (Boissonnade et al, 1985)

Finally, extensive validation with existing data is underway to calibrate the values and weights with the ranges assigned to the final answer. This way, the system can provide consultation comparable to an expert's performance.

Acknowledgements

The authors wish to express their appreciation to Professor James Gere and graduate students of Stanford's Department of Civil Engineering who contributed to the expert knowledge embodied in the system described in this paper. Partial support was attained from the National Science Foundation Grant No. CEE-8403516.

BIBLIOGRAPHY

Applied Technology Council, "Tentative Provisions for the Development of Seismic Regulations for Buildings," *ATC*, vol. 3-06, ATC, June 1978.

Boissonnade, A. C., Dong, W. M., Shah, H. C., and F. S. Wong, "Identification of Fuzzy Systems in Civil Engineering," *Proceedings of the International Symposium on Fuzzy Mathematics in Earthquake Engineering*, 1985.

Campbell, A., and T. Glover, *The Deciding Factor User's Manual*, Power Up Software, San Mateo, California, 1985.

Campbell, A. N., Hollister, V. F., Duda, R. O., and Hart, P. E., "Recognition of a Hidden Mineral Deposit by an Artificial Intelligence Program," *Science*, vol. 217, No. 3, Sept. 1982.

Duda, R., J. Gaschnig and P. Hart, "Model Design in the Prospector Consultant System for Mineral Exploration," *Expert systems in the Micro-electronic Age*, pp. 153-167, Edinburgh University Press, 1979.

Ishizuka, M., Fu, K. S., and Yao, J. T. P., "SPERIL-I: Computer Based Structural Damage Assessment System," *Report No. CE-STR-81-36, School of Civil Engineering, Purdue University*, 1981.

Ogawa, H., Fu, K. S., Yao, J. T. P. , "SPERIL-II -- An Expert System for Damage Assessment of Existing Structures," *Report No. CE-STR-84-11, School of Civil Engineering, Purdue University*, 1984.

Design of a Knowledge Based System
to Convert Airframe Geometric Models to Structural Models

Brent L. Gregory[1]
and
Mark S. Shephard[2]

INTRODUCTION

Finite element analysis techniques are commonly used to assess the safety and efficiency of aircraft structures. Ideally, these analysis would be carried out as an integral part of the design process. However, they are commonly only used in a post design phase to evaluate a final design which will be modified only if the results are unacceptable. This limited use of finite element analysis is not due to the computational cost required for the analysis, but is due to the time and cost required to generate the analysis models. Even when the geometry is available in a computer aided design (CAD) system, the available finite element modeling tools can not efficiently convert it into an appropriate numerical analysis model. This is because the complex nature of an airframe requires that the analysis model represent three-dimensional geometric components in terms of the appropriate combinations of one and two-dimensional structural elements. Current finite element preprocessors are designed for the efficient conversion of geometric entities to finite element models when the dimensionality of the two are the same; they do not efficiently handle the case where the dimensionality of the finite elements are lower than the geometric entities they represent.

As a result of this deficiency, airframe manufactures who use CAD tools to design airframes and finite element analysis software to analyze them, still generate the finite element models using semi-manual methods. This process works because the people creating the FEM understand the structural role of each airframe component and are able to capture its significance to the entire structure with a limited number of discrete elements. Often, the simplifications involve modeling three-dimensional structural components with a combination of one and two-dimensional finite elements, a technique beyond the capability of current automatic mesh generators. Thus, by using not only the geometric definition of each airframe component, but expertise on how to model its function, the analyst is able to create an accurate model simple enough

---

[1]   Research Assistant, Center for Interactive Graphics, Rensselaer Polytechnic Institute, Troy, NY 12180-3590.
[2]   Associate Professor of Civil Engineering, Rensselaer Polytechnic, Troy, NY 12180-3590.

to be analyzed using commercially available finite element modeling systems on large mainframes.

The drawbacks of the current method are obvious. Although the analysis can be carried out in one day, the semi-manual model generation process takes several months (1). While the geometric model is developed on the computer, and the final model is analyzed on the computer, the analysis model generation proceeds primarily by hand. Because of the time required for the generation of the analysis model, the analysis is used primarily for verification rather than for feedback to facilitate iterative design improvement.

The purpose of this paper is to present an approach by which:

• The entire modeling process is carried out on the computer using a combination of interactive and automated modeling procedures.

• Expert knowledge is used in the modeling software to increase the level of automation and to aid the analyst in making better modeling decisions.

• The analyst is relieved of the necessity of doing rudimentary calculations.

• The analyst can ask "why" the programmed modeling rules want to model a geometric component a certain way in the finite element model.

The result will be drastic reductions in the time and effort required, and improved reliability in the modeling process.

Basic Approach

The system described herein is called the "Flexible Automatic Conversion System (or FACS for short). The approach used in FACS is to give the computer access to the extra information, beyond the geometric airframe description, that will guide and aid the user in the creation of useful analysis models. This extra information takes the form of company analysis modeling manuals and expert knowledge. The manuals provide rules and guidelines to the engineer on how certain types of components can best be modeled in specific situations. On the computer, this information will be condensed into a set of rules telling how to make decisions about good simplification methods. To convert the geometry into an analysis model, the computer would need only to follow the rules. Modern artificial intelligence research provides a tool which can make the high-level modeling decisions by following a set of rules. A process which applies a set of rules to a particular situation is called an inference engine. If the rules correctly capture all the knowledge used by the airframe analyst expert then the inference engine will be able to function just as well as its human counterpart. (Hence the term, expert system.)

Creating the proper environment

The rules that are found in the analysis handbooks deal with the airframe at a higher level of abstraction than is normally accessible to the computer. Where the manual might talk about "frames", "cutouts" and the "skin", the computer deals primarily with geometric primitives like the coordinates of points and parametric equations of arcs and surfaces. To allow the inference engine to deal in the language of these abstractions, it must be able to access the geometric model in these high-level terms.

One approach to this problem is to use a geometric modeler which is designed specifically for airframes. Such a modeler would understand the components of airframes and would ask for and store the geometric data in terms of its place in the airframe structure. Thus, the inference engine could access the geometric model and directly retrieve information like structural component types, dimensions and other high-level parameters, since such abstract data would be an integral part of the special purpose airframe modeler's data base. This solution seems promising, but has some practical drawbacks. First, a special purpose airframe modeler would invariably be less flexible than than a general purpose three dimensional CAD system. It would always be difficult to model unusual airframe components if they did not conform to the special purpose modeler's specific data structure. A prototype special purpose airframe modeler has be created at RPI (3), but it was viewed as too restrictive for the airframe designer. This method would also require the airframe designer to learn a entirely new CAD system and would make his new airframe models incompatible with any preexisting CAD support systems. Thus, to avoid the mammoth task of creating a new geometric modeler, and to avoid incompatibility with existing systems it was decided to take the geometric models from an existing CAD system. Ideally, the geometric modeler, the FEM generating system and the analysis software would be integrated into a single coherent airframe development system. One which both the modeler and analysis system are geared especially to airframes, are at least as powerful as existing tools and are much more closely linked together. Such a system, though a worthwhile goal, is not only beyond the scope of our resources, but also those of the majority of airframe companies.

Since currently available capabilities require that analysis models be produced from geometric models stored in general purpose CAD systems, FACS was designed to process the low-level information found in their data bases. The problem of providing the inference engine with an environment where its rules, which operate on abstract objects, may be processed falls to a component of FACS called the Classifier. Generally, the Classifier takes the detailed and mathematical description of the airframe geometry and classifies it in the abstract terms used by the inference engine. The Classifier allows FACS to be used with commercial CAD systems currently being used for airframe design. Furthermore, it allows the rules to be written in a language similar to that used in the engineering handbooks.

COMPONENTS OF SYSTEM

Figure 1 depicts a data flow diagram of the components of FACS. Basically, FACS works as follows:

- First, the geometry extractor retrieves the geometric definition of the airframe from the geometric modeler's data base and breaks it up into its structural components.

- For each component of the airframe in the geometric model, the Classifier decides exactly what type of component it is, and extracts the require dimensional information from the geometric model.

- The Inference Engine then looks at the type of airframe component and other low-level information extracted by the classifier and based on a predefined set of rules supplied by the Rule Maintenance System, it chooses a method by which to model the component. It also decides certain conversion parameters such as the finenesses of the finite element meshes.

- Next, an Application Routine which carries out the chosen conversion method is invoked. It is supplied a template containing the conversion parameters dictated by the inference engine along with the geometric specification of the component. The Application Routines produces generic finite elements.

- Finally, A FEM translator is invoked to convert the generic finite elements into the syntax required by the target analysis package.

What follows is a description of the function of these entities.

Geometry Extractor

The function of the Geometry Extractor is to insulate the rest of the system from the particular CAD system from which the geometric model originates. A separate geometry extractor needs to be implemented for each CAD system that will be supported by FACS. The geometry extractor is the entity which knows how to access the data base of the source CAD system. It converts the geometry into a single generic format used by the rest of the system. The generic geometric model is independent of the particular geometric modeler in use.

Classifier

As mentioned before, the classifier must put the geometry into a meaningful form so that rules can be applied. To do this, the Classifier takes

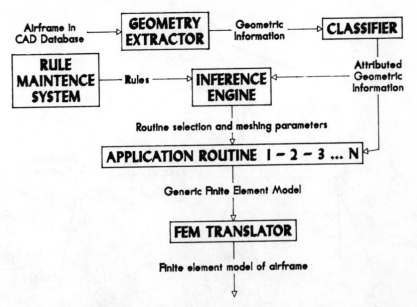

Figure 1. High level data flow through FACS

the geometry, which consists of geometric primitives such as points, lines, surface patches, and classifies them into groups of entities that define components of the airframe. The classification is the abstract name the expert uses to refer to the actual component geometry represents. Along with the classification, are the abstract parameters the expert uses to describe the component.

The classifier will be able to classify any geometry within the scope of the supplied classification system and the geometric primitives that are supported. The classification system is the guide by which the classification proceeds. It consists primarily of a classification tree. The root of the tree symbolizes all possible classifications, while the son of any given node represents an instanciation of the concept represented by that node. The leaves of the classification tree correspond to actual classifications suitable for use by the rest of the system. The process of classification then becomes the process of searching the classification tree for an appropriate classification leaf. Figure 2 depicts a simple example of a classification tree.

The appropriateness of a given classification is gauged by whether a particular piece of geometry can supply all the data required by the classification, as well as conforming to conditions laid out by the classification, such as the numbers of kinds of geometric primitives in a component. Each classification has associated with it a set of parameters which most be given values. For example, a certain "frame" classification might require a "station" value, two curves delineating the inside and outside of the frame, and a set of circles representing

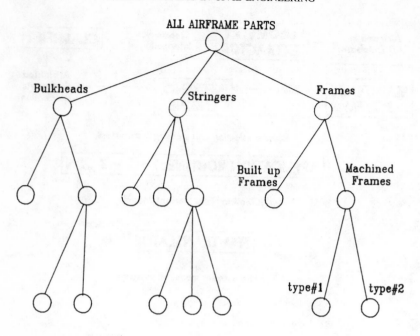

Figure 2. Classification Tree Example

cut-outs. A piece of geometry would be classifiable as this type of
"Frame" only if it could supply such information. Additional criterion
such as dimensional limits and restrictions on what other geometry is
present would also have to be satisfied before a particular classification could be accepted.

The process of classification can be thought of as a search for a leaf in
the classification tree which is a valid classification. The searching
for a classification is speeded by criteria at each node in the classification tree which tell if descendents of the node are worth searching.
The criteria at a non-leaf node are specified so that a piece of geometry
will be classifiable as one of the descendents of the node only if it
satisfies the criterion at that node. Satisfying the criterion, does
not, however, guarantee that a viable classification will be found among
the descendents. Thus, the purpose of non-leaf criterion is only to
expedite the search for an appropriate classification by eliminating
fruitless searching down the tree.

The inference engine.

The purpose of the inference engine to to decide which modeling method should be used to create the finite element representation of a given airframe component. It must further dictate the to the procedure which implements the actual conversion, factors which affects the degree fineness of the descretizations used in the chosen method. These decisions are made based on a system of rules supplied by an expert airframe analyst. The rules are supplied in the form: IF predicate THEN result. The predicates are Boolean formulas based on:

* Information about the airframe component supplied by the Classifier
* Information supplied by the results of other rules
* Known facts supplied with the rules such as data tables and defaults
* User responses to queries

Results can be:

* Directives telling which conversion method to use
* Directives telling meshing parameters
* New information on which other rules are based
    - New facts that are inferred about the airframe component
    - Directives telling when to seek user input or rely on defaults
    - Directives telling when to stop processing rules; either because all required information has been inferred, or because it cannot be inferred.

The inference engine has two methods by which to arrive at conclusions from the supplied rules. First, it searches for rules whose predicate can be satisfied. Once a "satisfied" rule is found, its result is processed, providing new information which can cause other rules to be satisfiable. This scheme is called forward-chaining. The second method begins with a hypothesis and tries to prove it by using rules whose result confirms the hypothesis and whose predicate is either a given fact or can be proved using other rules. This is called backward-chaining.

In general, backward chaining is most useful for arriving at a specific conclusions such as a choice of an application routine or meshing parameter. Since making such conclusions are the purpose of the inference engine in FACS, backward chaining is the key inference technique for our application. Backward chaining alone is very inefficient though, since the search for rules to satisfy hypotheses usually include many rules that do not apply. Forward chaining provides a means to limit the search to rules which conform to the known facts. Forward chaining by itself is also inefficient also, since the generation of new results is not directed toward any specific goal. The best inference engine algorithms use a combination of forward and backward chaining.

Rule Maintenance System

The rules which govern how the conversion is to take place must be developed by an airframe analysis expert. As the use of the system progresses, the rules will have to be expanded and refined to account for new types of airframe components and better handle existing ones. The Rule Maintenance System (RMS) provides a user friendly interface which facilitates the maintenance of the system of rules. The RMS must insure that the rules do not conflict with each other and that they provide enough information for the conversion to proceed.

Systems containing a RMS and an inference engine tools are commercially available (2).

Application Routines

The inference engine will not actually carry out the conversions of the geometric information to finite elements. It only decides how the conversion should take place. The job of actually creating the finite element model falls to the application routines. The application routines each implements a certain conversion methodology. The rules specify which of the routines are applied and how the conversion parameters are to be set.

Finite element translator

Like the geometry extractor(s), the FEM translator(s) insulate the rest of the system from the choice of a particular target finite element analysis package. The application routines produce an internal generic representation of the FEM which is independent of the particular analysis package being used. The FEM translator formats the generic FEM description into the syntax expected by the analysis package.

USE OF AUTOMATED DEDUCTION IN THE SYSTEM

Central to the design of FACS is the inference engine which is the replacement for the human expert airframe analyst. Given the rule-set, the inference engine is supposed to behave as the expert when selecting the conversion methods. While gaining the speed and accuracy, this replacement is far from perfect. Some of the considerations against using an expert system to do the thinking are:

- The inference engine cannot automatically adapt to new situations. If the character of the airframes that the system is to model change, the rules may have to be reworked.

- The inference engine cannot handle unusual cases well. Geometry may be discarded or poorly modeled if not accounted for in the rules.

- The entire expert system approach can be cumbersome. The inference engine and its supporting systems are all large complicated pieces of software which are difficult to integrate with the rest of the system. Unlike many expert system applications, this one must run in concert with a geometric modeler, a FEM analysis package and a large body of special purpose code.

- The extraction of expertise is both difficult and imperfect. The expert may not be able to describe his expertise easily as rules. Intuition, for example, is something that can not be digitized.

- Care must be taken to insure that the rules are consistent and that processing them will result in the desired answers.

Still, the reasoning power provided by the inference engine is indispensable and makes all the drawbacks necessary evils.

Deciding how much to rely on the inference engine, was a central factor in the design of the system. The problem was to minimize the cost of its use while still maintaining the required level of intelligence. The relationship between the classifier and inference engine exemplifies this trade off.

The classifier - inference engine interface

Central to the workings of FACS is the communication and work sharing that takes place between the classifier and the inference engine. If the system is to work, the inference engine must be able to test all the predicates of the rules either from information produced in other rules or from information produced by the classifier. So, for example, a rule which states: "IF the bulkhead is in the main cabin section THEN truss elements for flanges and shear panels for the webs should be used" would require the inference engine to be able to know whether the geometry was in the tail section or not before it could process the rule. If the rule system had rules which could tell whether the bulkhead was in the tail or not, then the inference engine would be able to use those rules, but if they didn't, then that information must have been presented by the Classifier. That is, one of the Classification parameters for the bulkhead must have been whether the geometry was in the main cabin or not.

This example illustrates that there is a division of labor between the classifier and the inference engine. The Classifier provides the lower level information tied closely to the geometric specifics of a given part (through parameter filling), while the inference engine can provide much more abstract and intelligent information. The decision of what the Classifier is to figure out and what the inference engine it to do rests in the hands of the people who set up the classification system and the rule base. What is required, though, is that all information required by rules be provided somewhere (either other rules or the Classifier).

It is conceivable that classification could be entirely carried out by an Inference Engine (in addition to the Inference Engine which decides how to perform the conversion). Where the Inference Engine we have already talked about deals in terms of abstract airframe concepts, the classifier Inference Engine's rules could deal with a geometric data such as the primitive type and coefficients. While this would undoubtedly make the classifier more powerful, this approach is currently not being pursued in the development of FACS. This is because it is assumed that classification can be performed by reasonably straight forward algorithms so the use of an inference engine to do its work would only make it less efficient. The designers of FACS felt that an expert system approach should only be taken when standard algorithmic approaches fail. This is because while the inference engine is capable of solving complex problems, it does do at the cost of being large and cumbersome and relatively inefficient. It would be possible, for example, to implement amortization computations with an expert system, but it is not necessary since simple efficient algorithms can be written to perform this function.

User Queries

Just as the inference engine makes the modeling decisions which are too complex for the classifier to handle, the user may have to do the thinking for the inference engine when things get too complicated. Ideally, classification and rule processing would always proceed automatically. In reality, however, there will always be the possibility of unexpected geometry for which the rule or classification systems failed to account. When such a situation occurs, the system will consider three alternatives; resort to default values, discard the portion of the geometry which caused the problem, query the user for the answer.

The inference engine uses the rule-set to determine which course of action to take. Built into the rule-set are rules which will reach conclusions based on user input; will tell the inference engine when it can give up and just dump some geometry; will decide when needed information can be taken from default tables.

One of the bonuses that comes with the use of an inference engine is the ability of the engineer to ask "why" a query is made when user input is sought. The inference engine can explain its reasoning process by detailing the rules it is using and the information it requires to process those rules. This way, the engineer is able to make better modeling decisions by a closer understanding of the modeling process.

The classifier will be faced with dilemmas whenever geometry does not fit into the the classification system or can be classified more than one way. In such cases, the action of the classifier can be dictated by run-time parameters. That is, whether the geometry is discarded, given a default classification or a user query is instituted will depend on parameters set when FACS is invoked.

Control over how dilemmas are resolved is left in the hands of the user. By setting run-time parameters and by changing the rule-set, the user of FACS may determine exactly how automatic the conversion will be.

The cost of more automation could be unmodeled or poorly modeled geometry, while the cost of user queries is an increase in modeling time and a loss in model consistency. It is up to the user to strike a balance between these two.

Still, the rule-set can always be improved to automatically account for geometry which previously caused uncertainty. This can be done by codifying the expertise used by the person running the program to answer the inference engines queries. Classification dilemmas and be avoided by improving the classification tree. Extra criterion could be added to distinguish between ambiguous classifications. Also, new classifications could be added to handle previously unclassifiable geometry.

STATUS OF SYSTEM DEVELOPMENT

A prototype of FACS is now under development at RPI. The prototype will run on an IBM minicomputer running the VM/CMS operating system. The geometric models will originate from the CATIA geometric modeler. The prototype will generate generic models which can be directly converted to the format expected by standard finite element analysis packages. The geometry extractor and classifier are currently being coded in PL/1. PRISM, an expert system framework will provide a rule maintenance facility as well as the inference engine.

The first version of the Classifier is being designed so that it will take advantage of "hints" in the geometry database. These hints give it a head start by telling a node in the classification tree which is a parent of the actual classification. The hints serve to make classification more accurate and efficient by limiting the scanning to a small subset of the whole classification tree. The hints are entered into the database using a capability of CATIA to associate a character string with a geometric element. CATIA also has the capability of grouping related geometry into sets. The current prototype relies on the designer to segregate the geometry of each airframe component into its own set. While the inclusions of classification hints is optional in the prototype, the grouping into sets is required.

It is the goal of FACS to minimize the requirements made on format of the geometry inside the geometric model. The requirements defined so far specify techniques that are currently being used, so little changes should be required to use the FACS. Invariably, however, the efficiency of the analysis model generation process will require some formalization in the geometric model creation.

CONCLUSIONS

The automation of airframe analysis is a problem which stretches the limits of current software capabilities. To address it, FACS combines the use of standard algorithmic programming with expert systems technology. By using an inference engine as a tool to perform highly abstract "intelligent" decision making, FACS will be able to efficiently generate the quality of model previously attainable only by manual methods. Standard algorithmic programming is used extensively to transform the problem into an abstract domain best suited to the inference engine and to perform most of the calculations required by the conversion.

The anticipated result will be a near-total automatization of the airframe analysis process. This should drastically improve the way airframes are developed. By making analysis cheaper and faster, it will be more accessible earlier in the development process. This in turn should improve airframe designs by detecting problems earlier in the development cycle.

ACKNOWLEDGEMENT

This work is jointly funded by the Army Research Office (Contract Number DAAG 29-82-K-0093), Dr. Robert Singleton, project manager, through the Center of Excellence in Rotocraft Technology at Rensselaer Polytechnic Institute, and by the Industrial Associates of the Center for Interactive Computer Graphics at RPI.

REFERENCES

1. R. Gabel, W. J. Kesack and D. A. Reed, "Planning, Creating and Documenting a NASTRAN Finite Element Vibrations Model of a Modern Helicopter", NASA Contractor Report, Boeing Vertol Company, 1982.

2. Harmon, P. and King, D., Expert Systems, John Wiley and Sons, Inc., New York, New York, 1985.

3. B. P. Johnston and M. S. Shephard "Interactive Computer Modeling of Airframe Structures", Journal of American Helicopter Society, Vol. 30, No. 3, 1985, pp 59-61.

FAULT DIAGNOSIS OF HAZARDOUS WASTE INCINERATION FACILITIES
USING A FUZZY EXPERT SYSTEM

Y.W. Huang, S. Shenoi, A.P. Mathews[*], F.S. Lai[**], and L.T. Fan

Institute for System Design and Optimization
College of Engineering
Kansas State University
Manhattan, Kansas 66506

ABSTRACT

The use of an expert system along with fuzzy fault tree analysis is discussed in connection with fault diagnosis of a hazardous waste incineration facility. This fault tree is a diagrammatic representation of knowledge about the modes of failure of the incineration facility, and is readily transformed into a collection of production or IF-THEN rules. By incorporating the notion of fuzzy probabilities into fault tree analysis, the expert system is capable of drawing meaningful conclusions from uncertain situations. This greatly enhances the flexibility of the program.

The present work utilizes M.1, a commercially available expert system shell, in building the expert system. The fault tree for the hazardous waste incineration system, provided by a domain expert, is represented in the form of IF-THEN rules in the shell. Examples of fault detection are presented.

*   To whom correspondence should be addressed.
** USDA-ARS, U.S. Grain Marketing Research Lab., Manhattan, KS66502

INTRODUCTION

Over the last decade, our society has become much more concerned about the treatment of hazardous wastes. In reaction to this awareness, the Resource Conservation and Recovery Act of 1976 (RCRA) and its 1980 regulations were enacted. RCRA has been designed to guarantee that all large quantities of hazardous waste will be routed through approved storage, transportation, treatment, and disposal facilities. The recently enacted Hazardous Waste Act Amendments of 1984 and the new evolving RCRA regulations place stringent controls on the ultimate disposal of hazardous wastes. As a result, the quantity of waste that will be incinerated, and the number of operations involved will increase dramatically in the future. Nevertheless, since failure in the incineration and air pollution control systems could have devastating effects, it is important to analyze all possible failure modes and to design countermeasures and diagnostic systems for quickly locating and dealing with sources of failure. This paper discusses the use of an expert system along with fuzzy fault tree analysis for diagnosing malfunctions in hazardous waste incineration systems.

Hazardous wastes, as defined by the Environmental Protection Agency, include (1) ignitable waste, Hazard Code "I"; (2) corrosive waste, Hazard Code "C"; (3) reactive waste, Hazard Code "R"; (4) extract procedure toxic waste, Hazard "E"; (5) acute hazardous waste, Hazard Code "H"; and (6) toxic waste, Hazard Code "T". As a result of an increased public awareness of the effects of these wastes on the environment, and the quality of our natural resources and national health, the Federal Government has established statutory control over discharges into the environment. Moreover, as energy costs increase and the quality of wastes changes with advances in technology, many so-called wastes are no longer discarded but are exploited as fuels. Consequently, the use of incineration facilities in hazardous waste treatment has attracted increased research attention. A discussion on the applicability of incineration techniques can be found in Brunner, (1984).

An expert system is a computerized problem-solver containing a large body of knowledge in a restricted domain; it uses this knowledge to draw conclusions by emulating the performance of human experts. Stripping the knowledge component of the expert system yields a shell containing only the linguistic and reasoning mechanisms; these mechanisms interpret and draw conclusions when knowledge is provided (Barr and Feigenbaum, 1981). The shell employed in this study is the microcomputer-based M.1; it embodies knowledge in the form of rules and employs a backward chaining or goal-driven control strategy.

A fault tree is a diagrammatic representation of knowledge about the modes of failure of a system. The information contained in the fault tree is readily transformed into a collection of production rules or IF-THEN rules. In conventional fault tree analysis, it is necessary to assign numerical values for event probabilities; the values are generally not available for events with limited frequencies of occurrences. By incorporating the notion of fuzzy probabilities into fault tree analysis, it is possible to develop a user-oriented and highly transparent expert system for fault detection.

FUZZY FAULT TREE ANALYSIS

A fault tree (see, e.g., Haasl, 1965; Barlow and Chatterjee, 1973; Barlow and Lambert, 1975) is a logical and hierarchical description of an accident (top event) in terms of sequences and combinations of malfunctions of individual components and adverse operating conditions (basic or fundamental events). Figure 1 depicts a simple fault tree. The construction of a fault tree is a major task in hazard analysis. By resorting to fault tree analysis, the reliability of a complex system, such as the hazardous waste incineration facilitiy under consideration, can be computed in terms of the probabilities of occurrence of the basic events. In conventional fault tree analysis, probabilities are assigned to the basic events of the system under consideration. These probabilities are combined according to the structure of the AND-OR fault tree.

In many instances, e.g., hazardous waste incineration, it is difficult to estimate exactly the failure rates of individual components or the probabilities of occurrence of undesirable events. Some extremely hazardous accidents may never have occurred before, or occur so rarely that sufficient or meaningful statistical data are not available. To cope with the problem, it is appropriate to employ the concept of "fuzzy probabilities" (Noma et al., 1981; Tanaka et al., 1983) in articulating human subjective notions of probabilities. A fuzzy probability can be viewed as a fuzzy set defined on a probability space, expressing the subjective notion that the probability of occurrence of an event is approximately equal to a certain value.

The motivation behind the use of fuzzy probabilities for dealing with hazardous events can be discerned from a common situation in the real world. When asked for an estimate, an experienced operator might be unwilling to state that the probability of excessive feed rate to a waste incinerator is exactly equal to 0.5; instead, he would definitely feel more comfortable to express the probability of occurrence as a range of values, e.g., "around 0.5". Thus, the chance that the feed rate is excessive could be as high as 0.75, or as low as 0.25. A fuzzy probability of "around 0.5" is illustrated in Figure 2, in which Figure 2(a) is a common convex fuzzy set (see e.g., Zimmermann, 1985) and Figure 2(b) is a trapezoidal representation of the same concept (see e.g., Lai et al., 1986). The fuzzy probability of occurrence of a high feed rate can be denoted by a four-tuple

$$P(X_i) = [q_i^l, p_i^l, p_i^r, q_i^r]. \tag{1}$$

The trapezoidal representation corresponding to the fuzzy probability of "around 0.5" for an excessive feed rate to the incinerator could be expressed as

$$P(high\_feed\_rate) = [0.25, 0.4, 0.6, 0.75]. \tag{2}$$

To propagate fuzzy probabilities from the basic events to the top event, Tanaka et al. (1983) have introduced a multiplication operator (⊗) for trapezoidal fuzzy sets.

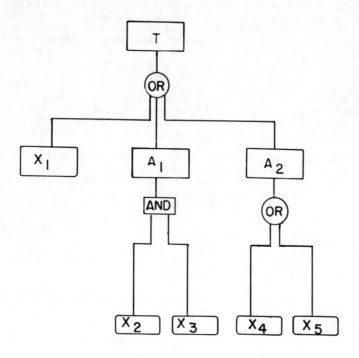

Figure I. An example of a fault tree.

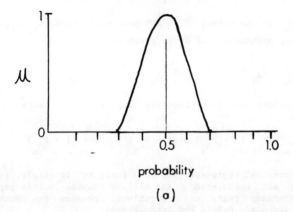

probability

(a)

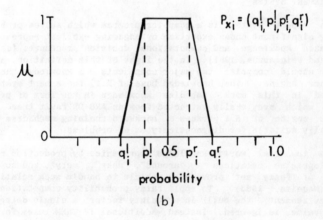

$P_{x_i} = (q_i^l, p_i^l, p_i^r, q_i^r)$

probability

(b)

Figure 2. Representation of the fuzzy concept "around 0.5".

$$P(X_i) \bullet P(X_j) = [q_i^1 q_j^1, \; p_i^1 p_j^1, \; p_i^r p_j^r, \; q_i^r q_j^r] \qquad (3)$$

If event A_m is the output of an AND gate with inputs A_1, A_2, ..., A_n, then the fuzzy probability of A_m is given by

$$P_{A_m} = P_{A_1} \bullet P_{A_2} \bullet \ldots \bullet P_{A_n} \qquad (4)$$

The corresponding fuzzy probability for an OR gate with n inputs is expressed as

$$P_{A_m} = 1 - (1 - P_{A_1}) \bullet (1 - P_{A_2}) \bullet \ldots \bullet (1 - P_{A_n}) \qquad (5)$$

The trapezoidal representation of a fuzzy set is simple; it facilitates fuzzy set computation, and will be adopted in this paper. For details concerning fuzzy set operations, readers can consult other references (see e.g., Dubois and Prade, 1980).

FUZZY EXPERT SYSTEM

An expert system is a computer program which solves problems requiring significant human expertise by engaging explicit representations of domain knowledge and computational decision procedures (see, e.g., Barr and Feigenbaum, 1981). In the light of this definition, an expert system should comprise two major components - a knowledge base and an inference engine. The knowledge base of M.1, the expert system shell employed in this work, embodies knowledge in the form of production rules, which are easily extracted from an AND-OR fault tree. The inference engine of M.1 pursues a backward chaining mechanism which is especially suitable for diagnostic types of problems.

In the real world, knowledge represented by production rules can have imprecise relations - "around", "near", "very", and so on. The notion of fuzzy set provides a vehicle to handle such relations (see e.g., Negoita, 1985). To apply fuzzy probability computations in the M.1 environment, the built-in certainty factor, a single deterministic belief value, is ignored. Instead, additional IF-THEN rules for manipulating fuzzy probabilities via the appropriate fuzzy set operations are incorporated in the knowledge base. Various schemes have been suggested for acquiring fuzzy probabilities from domain experts (Zimmermann, 1985). In this paper, we employ two acquisition schemes; the methodologies are illustrated in an implementation example.

EXAMPLE

A block diagram of a typical incineration system is illustrated in Figure 3. Wastes fed into an incinerator are thermally decomposed via oxidation with the aid of fuel and combustion air. Two major factors affecting the system performance are the temperature in the incinerator and residence time of the wastes through it. A fault tree describing the possible failure modes of the system is presented in Figure 4.

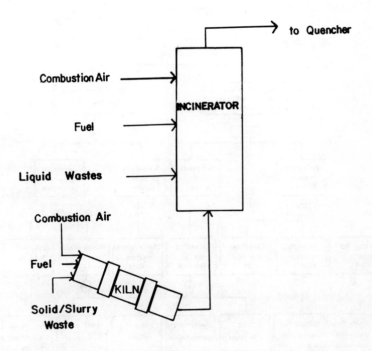

Figure 3. A block. diagram of a typical
incineration system.

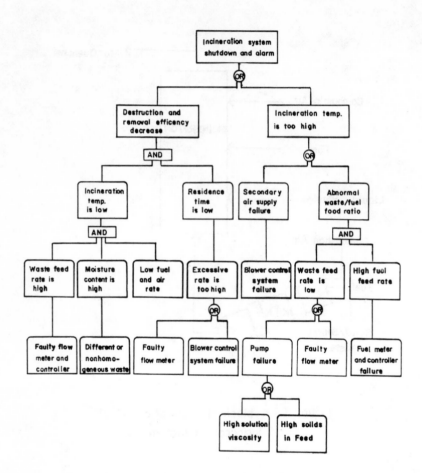

Figure 4. Partial fault tree for hazardous
waste incineration facilities.

The top event of the tree in Figure 4 corresponds to a malfunction of the incineration system which may occur when either the destruction and removal efficiency of wastes decrease or the incineration temperature is excessively high or both. A decrease in the destruction and removal efficiency of wastes will only occur provided that both the incineration temperature and the residence time are low. A fault tree expressing the occurrence of the top event in terms of various basic events can be constructed through an operability study (see e.g., Lawley, 1974). The resultant fault tree representation is easily transformed into production rules. For instance, a rule concerning the waste destruction and removal efficiency can be written in the form

> if incineration_temp_low and
> residence_time_low
> then destruction_and_removal_efficiency_decrease.

Except for rare situations, however, we may find it extremely difficult to draw a precise line between the concepts of low and high temperatures. Fuzzy probability provides a powerful tool for circumventing this difficulty. In the present study, two schemes are employed to acquire fuzzy probabilities.

In the first, rules are devised to accept fuzzy probabilities directly from an operator. For example,

> rule-1: if X(Y) is yes
> then fuzzy_prob(Y) = [1.0, 1.0, 1.0, 1.0].

> rule-2: if X(Y) is no
> then fuzzy_prob(Y) = [0.0, 0.0, 0.0, 0.0].

> rule-3: if not (X(Y) is yes) and
> not (X(Y) is no) and
> estimation (Y) is [A, B, C, D]
> then fuzzy_prob(Y) = [A, B, C, D].

where X(Y) is the operator's response concerning the occurrence of a basic event "Y". If X(Y) is yes, then event "Y" is positively confirmed to have occurred; hence, the fuzzy probability of occurrence of event "Y", fuzzy_prob(Y) is defined to be [1.0, 1.0, 1.0, 1.0]. Similarly, if the response to X(Y) is no, then event "Y" is negatively confirmed and fuzzy_prob(Y) is defined to be [0.0, 0.0, 0.0, 0.0]. When the operator is unsure about the situation, [the response to X(Y) is neither "yes" nor "no"], the program will ask for a fuzzy estimate of the probability based on the trapezoidal representation. The fuzzy probabilities for the basic events obtained from the operator will then propagate in accordance with Equations (3) through (5). A snapshot of the program at run-time is given in Figure 5.

Although a distinct line between "true" or "false" of a statement is not clear in some cases, the upper and lower bounds are quite certain. For example, in spite of not knowing the exact value that is high for residence time, the operator of an incinerator may agree that a residence time higher than 2.0 seconds is certainly too high, while a value lower than 0.5 second is definitely too low. A residence time

What is the residence time (in seconds) of waste in the incinerator?
(Enter the time or a "0" for unknown)

>> 0.

Even you do not know the precise residence time, give me a time range (in seconds) in the form of [upper_bound, lower_bound].

>> [0.5, 0.9].

The residence time of the waste in the incinerator is fuzzily low ({ 0.638, 0.752, 0.981, 1.0}).

Is there any faulty_flow_meter?

>> yes.

Is there any faulty_flow_controller?

>> not sure.

Even faulty_flow_controller is uncertain, try to fuzzily describe the situation using trapezodial fuzzy set representation.

>> [0.25, 0.3, 0.4, 0.5].

Waste feed rate is high to a fuzzy degree of { 0.25, 0.3, 0.4, 0.5 }.

Is there any different_types_of_waste?

>> yes.

The moisture content is too high.

Is there any low_fuel_rate?

>> yes.

Figure 5.   Example of the expert system at run time

Is there any low_air_rate?

>> maybe.

Even low_air_rate is uncertain, try to fuzzily describe the situation using trapezodial fuzzy set representation.

>> [0.5, 0.55, 0.6, 0.65].

Fuel and air rate are low to a fuzzy degree of (0.125, 0.165, 0.24, 0.325).

Incineration temperature is low to a fuzzy degree of (0.125, 0.165, 0.24, 0.325).

Destruction and removal efficiency is fuzzily ((0.0798, 0.124, 0.235, 0.325]) decreasing.

. . . .

Figure 5 (continued).

between 0.5 and 2.0 seconds is then considered "fuzzily" high or low. This notion is employed in the second scheme for manipulating fuzzy probabilities.

In the second scheme, certain facts are first installed in the data base. For the case involving the residence time, we have

    fact-1: upper_bound(residence_time) = 2.0.
    fact-2: lower_bound(residence_time) = 0.5.

When asked for an estimate of the residence time, an operator can either reply with a single value or a range of values. If the single value is provided, a fuzzy number will be derived based on the upper and lower bounds of the residence time. For example, a residence time of 0.4 second results in a fuzzy set of [1.0, 1.0, 1.0, 1.0] indicating that the residence time is unbearably low. A residence time of 3.0 seconds yields a fuzzy set of [0.0, 0.0, 0.0, 0.0] corresponding to a value that is certainly too high. On the other hand, if the residence time is 0.7 seconds, a fuzzy set of [0.87, 0.87, 0.87, 0.87] will be generated. The following rules are employed in implementing these operations.

    rule-9:   if upper_bound(X) = U and
                  lower_bound(X) = L and
                  U - L = F
               then fullscale(X) = F.

    rule-10-2: if not residence_time = 0 and
                  fullscale(residence_time) = F and
                  lower-bound(residence_time) = L and
                  residence_time = R and
                  (R - L)/F = P
               then rindex = P.

    rule-11-1: if rindex = P and P >= 1
               then b(residence_time) = [1.0, 1.0, 1.0, 1.0].

    rule-11-2: if rindex = P and P =< 0
               then b(residence_time) = [0.0, 0.0, 0.0, 0.0].

    rule-11-3: if not residence_time = 0 and
                  rindex = P and P < 1 and P > 0
               then b(residence_time) = [P, P, P, P].

where b(residence_time) is a fuzzy description of the probability that a residence time is too high. With further manipulation, a fuzzy prob- ability corresponding to too low a residence time can be obtained. To cope with situations in which the operator cannot provide an exact value for the residence time, additional rules are invoked for deriving a fuzzy probability from a range that may be supplied.

    rule-10-1: if residence_time = 0 and
                  your_guess(residence_time) = [G1, G2] and
                  fullscale(residence_time) = F and
                  (G1 + G2)/2 = A and
                  lower_bound(residence_bound) = L and

$$(A - L)/F = P$$
then rindex = P

rule-11-4: if residence_time = 0 and
rindex = P and P>1 and P<0 and
your_guess(residence_time) = [G1, G2] and
(G1 + G2)/2 = A and G2 - G1 = I and I/A = R and
$P - 0.4*R = P1$ and $P - 0.2*R = P2$ and
$P + 0.2*R = P3$ and $P + 0.4*R = P4$ and
then tempo(residence_time) = [P1, P2, P3, P4].

where rule-11-4 contains subjective subrules for generating a fuzzy set.
Observe that the greater the range of guessed values, the wider the
spread of the trapezoidal fuzzy set values. If the residence time is
expressed as the range [0.5, 0.9] based on these rules, then the
tempo(residence_time) will have a fuzzy set value of [-0.10, 0.02, 0.24,
0.36], which describes a degree of "highness" of the residence time. By
taking the complement of tempo(residence_time) and by forcing the fuzzy
set tuples to have values between zero and one, we obtain a fuzzy set
[0.64, 0.76, 0.98, 1.0] for denoting the fuzzy probability that the
residence time is too low. Note that the average value of the range
[0.5, 0.9] is 0.7; this is a situation discussed earlier. A snapshot of
the scheme is also illustrated in Figure 5.

CONCLUDING REMARKS

A rule-based expert system is a powerful vehicle for embodying and
engaging knowledge in diagnostic problem-solving. A fault tree is a
diagrammatic representation of the modes of failure of a system constructed through an operability study which is easily transformed into a
collection of production rules. The concept of fuzzy probability provides a viable approach for dealing with the subjective uncertainty
encountered in real-world situations. The present example involving
fault diagnosis in a hazardous waste incineration facility demonstrates
that there is much to be gained in jointly pursuing these three important concepts and methodologies.

ACKNOWLEDGEMENT

The authors gratefully acknowledge the partial financial support
provided by the National Science Foundation (Grant DMC-8516870).

REFERENCES

1. Barlow, R. E., and Chatterjee, P., "Introduction to Fault Tree Analysis," Operations Research Center, University of California, Berkeley, Report OCR 73-70, 1973.

2. Barlow, R. E., and Lambert, H. E., "Introduction to Fault Tree Analysis," in Reliability and Fault Tree Analysis - Theoretical and Applied Aspects of System Reliability and Safety Assessment, Barlow, R. E., Fussel, J. B., and Singpurwalla, N. D., Society for Industrial and Applied Mathematics, Philadelphia, 7, 1975.

3. Barr, A., and E. A. Feigenbaum, The Handbook of Artificial Intelligence, 1, William Kaufmann, Los Altos, CA, 1981.

4. Brunner, C. R., Incineration Systems - Selection and Design, Van Nostrans Reinhold, New York, New York, 1984.

5. Dubois, D., and Prade, H., Fuzzy Sets and Systems - Theory and Applications, Academic Press, New York, New York, 1980.

6. Haasl, D. F., "Advanced Concepts in Fault Tree Analysis," Proceedings System Safety Symposium, The Boeing Company, Seattle, WA, June 8-9, 1985.

7. Lawley, H. G., "Operability Study and Hazard Analysis," Chemical Engineering Processing, 55(4), 58 (1974).

8. Lai, F. S., Shenoi, S., and L. T. Fan, "Fuzzy Fault Tree Analysis: Theory and Application," To appear in Engineering Risk and Hazard Assessment, CRC Press, 1986.

9. Negoita, C. V., Expert Systems and Fuzzy Systems, Benjamin/Cummings, Menlo Park, California, 1985.

10. Noma, K., Tanaka, H., and Asai, K., "On Fault Tree Analysis with Fuzzy Probability," J. Ergonomics, 17, 291 (1981) (in Japanese).

11. Tanaka, H., Fan, L. T., Lai, F. S., and Toguchi, K., "Fault Tree Analysis by Fuzzy Probability," IEEE Trans. Reliability, R-32, 453 (1983).

12. Zimmermann, H. J., Fuzzy Set Theory - and Its Applications, Chap. 10, Kluwer-Nijhoff Publishing, Hingham, MA, 1985.

An Expert System for
Inactive Hazardous Waste Site Characterization

Kincho H. Law[1], AM ASCE

Thomas F. Zimmie[2], M ASCE

David R. Chapman[3], AM ASCE

Abstract

The area of hazardous waste management is a broad and multidisciplinary field requiring expertise in engineering, geology, chemistry and toxicology, and is an ideal area for the application of expert systems. This work deals specifically with inactive hazardous waste site investigations. OPS5, a production rule programming system, is used to implement the prototype expert system. The determination of permeability of the soil and the ground water flow direction and gradient are used as examples in this paper.

Background

The area of hazardous waste management is truly a broad and multidisciplinary field and an ideal area for the use of artificial intelligence and expert systems. This paper deals specifically with inactive hazardous waste site investigations. There are many thousands of inactive hazardous waste sites in the country, and the investigation and characterization of these sites is important, so that the risks to the environment and public health can be properly assessed, and the sites prioritized relative to possible remedial actions.

Hazardous waste site investigations require expertise in such fields as engineering, geology, chemistry and toxicology. Ideally, on a site investigation, an expert with knowledge in all the fields is available, or else a team of experts in each field is assembled. This approach is often feasible on large jobs with large budgets. Unfortunately, this does not occur on the typical hazardous waste site investigations, largely because of budgetary constraints.

Generally, there are several phases for investigations of inactive hazardous waste sites. The first phase, often referred to as a Phase I investigation, generally consists of a site visit, and a

---

[1] Assistant Professor, [2] Associate Professor, and [3] Graduate Student, Department of Civil Engineering, Rensselaer Polytechnic Institute Troy, New York 12180-3590

search of existing records and data on the site, but no subsurface
exploration, sampling or chemical analyses. Costs are on the order of
several thousand dollars.

The next phase, referred to as Phase II investigations by many
states, usually includes subsurface exploration, installation of
several monitoring wells, soil and water sampling, and chemical
analyses. Typical budgets range from $50,000-80,000. It is generally
the most difficult to obtain the proper expertise in the Phase I and
Phase II type investigations.

Regardless of the type of investigation, the site must be scored,
or ranked, relative to the risk posed by the site. The present system
used is the Mitre model, named after the company that developed the
system. Officially, it is known as the Hazard Ranking System (HRS)
and is used by the U.S. Environmental Protection Agency (EPA) and
state environmental agencies to evaluate the relative potential of
uncontrolled hazardous waste facilities to cause safety problems or
ecological and environmental damage [1]. In the HRS, the potential
hazards of land disposal facilities are categorized into three major
modes; (1) migration of pollutants via ground water, surface water and
air routes; (2) fire and explosion potential; (3) direct contact with
hazardous substances. Each of the hazard modes is subdivided into
several categories which contain a number of factors to be scored.
Although this system is easy to use, Wu and Hilger [2] point out
some drawbacks and misleading interpretations in real situations.

In fact, the object or goal of a site investigation is to
properly characterize the site relative to safety, ecological and
environmental risks. The Mitre model should perhaps be viewed as a
tool towards that aim. The final HRS numerical score assigned for a
site really is a summary of the information available on a site, and
allows for rank ordering of sites throughout the country. It is
recognized that individual site scores can have important political
and economic consequences for site owners and state regulating
agencies, since a sufficiently high score (presently 28.5 and greater)
allows the site to be listed on the National Priority List, and to be
considered for federal aid towards remedial actions. Nevertheless,
the detailed and thorough site information is found in the
documentation and records that supplement the use of the Mitre model.
That is, expertise is required to characterize the site, including the
appropriate documentation to support the conclusions; whereas
relatively little expertise is required to do the Mitre scoring after
the proper documentation exists.

Relative to expert systems, the actual Mitre scoring is a rather
trivial task, since the need for expertise is greatest in the area of
site characterization and documentation. On large projects, where
proper expertise is available, the use of an expert system can still
be valuable, at a minimum serving as a thorough checklist of items to
be considered.

One of the most basic pieces of information desired in a hazardous waste site investigation is the direction of ground water flow, and one of the most important site properties desired is the permeability (or hydraulic conductivity) of the soil or rock underlying the site. The determination of these items in the site characterization process are used as examples in this paper.

Production Rule Systems

An expert or knowledge-based system is generally referred to as a computer program that is able to perform a task within the task domain at the level of a human expert in that domain. Often, knowledge can be represented as facts and rules that are well suited for implementation in production system environment. A typical production system architecture consists of three major components [3]:

(1) a production memory storing knowledge facts and rules;
(2) a working memory representing the facts and assertions about the problem;
(3) an inference engine providing a control strategy in matching rules, selecting rules and executing rules.

Typically, a production system program consists of an unordered collection of production rules expressed in the form consisting of an "if" part and a "then" part [4]:

```
IF        condition 1
          condition 2
               .
               .

THEN      action 1
          action 2
               .
               .
```

The "if" part is usually a boolean combination of clauses and the "then" part specifies a set of actions modifying the working memory or performing operations. As an example, one production rule to determine ground water release information accoring to the Mitre model can be stated as:

```
IF        the depth of the aquifer is shallow (< 20 feet)
   and    the net precipitation is high      (< 15 inches)
   and    permeability is high               (< 1.0E-3 cm/sec)
   and    the hazardous substance is liquid, sludge or gas
THEN      a ground water release is virtually certain.
```

Rules in a production system program can be applied in either direction corresponding to the control strategy of the inference engine. In the forward chaining strategy, the conditions in a rule specify the combination of facts or objects to be matched against the current situation noted in the working memory. The backward chaining strategy uses the action part for matching, to hypothesize the goal or subgoals needed to be resolved.

The prototype expert system for hazardous waste site characterization described in this paper is developed based on a production
system language OPS5 [5]. Expert system programming in OPS5 has been
discussed in detail in a recent book [3]. The following briefly
describes the basic components in an OPS5 program.

An OPS5 program consists of a declaration section and a
production section. The declaration section describes the data
objects and their attributes permitted in the system. The declaration
of a data object "goal" with attributes "objective," "action" and
"status" would be declared as:

          (literalize goal objective action status)

The production section contains the knowledge facts expressed in
rules. For instance, the production rule:

          IF   the objective is (to determine) permeability
            and the action is input
            and the status is active
          THEN read soil-material and hydraulic-conductivity
                    (if known)
            and proceed to classify the permeability level

can be translated in OPS5 as

          (p permeability==input
             {(goal ^objective permeability
                    ^action input
                    ^status active) <goal>}
          -->
             (bind <soil> (accept))
             (bind <conductivity> (accept))
             (make permeability ^soil_material <soil>
                                 ^hyd_cond <conductivity>)
             (modify <goal> ^action classification))

The data are kept in working memory by objects. During execution, the
working memory will be modified by the action. In the above example,
the working memory

     (goal ^objective permeability ^action input ^status active)

becomes

     (goal ^objective permeability ^action classification
           ^status active)
     (permeability ^soil_material clay ^hyd_cond nil)

where "clay" and "nil" are input data by the user. It sould be noted
that a production system program is a data-driven program instead of

an instruction-driven program written in most procedural languages; that is, the communications among the rules are only by way of data. Although the inference engine in OPS5 supports only forward chaining, backward chaining strategy can also be implemented by splitting a goal into subgoals in a recursive manner. In addition, OPS5 allows external functions to be defined and called by the users. The external functions are particularly useful when numerical computations and algorithmic solution processes are necessary. In this work, external functions are implemented in COMMON LISP.

Knowledge Information

A production rule system is particularly suitable for loosely coupled problems, such as classification problems, that are decomposable into relatively independent subproblems. The hazardous waste site characterization problem involves many factors ranging from geological information to the chemistry of hazardous substances, each of which itself is a subtask which must be addressed. In the Mitre (HRS) system, evaluation of ground water migration is subdivided into four categories: route characteristics, containment, waste characteristics and the target (site environment). This work focuses on the ground water routing problem.

Knowledge information includes facts and rules appearing in handbooks as well as rule-of-thumbs that an expert would use in solving a particular problem. An expert system should also model the expert's problem solving strategy. For instance, the ground water information such as flow direction and gradient are of major concern in hazardous waste transport, but are not specifically required by the Mitre model. Soil stratification and water table elevation may influence the expert's decision in determining the permeability level at the site. This information should be incorporated in the expert system. This work attempts to include into the knowledge base the information provided by the HRS system as well as the information that an expert would use as a "check list" to characterize a hazardous waste site. The following discussion focuses on (1) determination of the permeability level and (2) determination of ground water flow direction and gradient.

Determination of Permeability Level

In the HRS system, permeability levels are classified into four categories: very low permeability (score 0), low permeability (score 1), moderate permeability (score 2), and high permeability (score 3). The scoring can be determined according to the numerical value of hydraulic conductivity, if available, or the soil material type. A typical rule can be stated as:

> IF   the objective is (to determine) permeability level
>   and the action is classification
>   and the hydraulic conductivity is unknown (nil)

and the soil material is one of clay, compact-till,
                    shale, or unfractured-rock
THEN   the permeability level is determined to be very low
            based on the given soil material
        and the scoring of 0 is assigned
        and classification is complete.

which can be translated as:

```
(p make-classification==very-low-permeability-material
    {<goal> (goal ^objective permeability
                   ^action classification ^status active)}
    {<perm> (permeability ^hyd_cond nil
                   ^soil_material << clay compact-till
                   shale unfractured-rock >>)}
    -->
    (modify <perm> ^level very-low ^source soil_material
            ^score 0)
    (modify <goal> ^action classification
            ^status complete))
```

When soil stratification is observed, however, the determination
of a permeability level is not as obvious.   Figure 1 illustrates the
situation of stratification for the cases of two soil layers with
different permeabilities.   In this situation, the water table becomes
a factor in the determination of the permeability level.   In general,
horizontal flow occurs at the same layer where the water table
resides.   However, vertical flow occurs until the water table is
reached.   In general, horizontal flow is undesirable, since it allows
contamination to spread and move off-site, whereas vertical flow
confines the contamination to the site.   Hence, if there is a less
permeable layer above the layer at the water table, the less permeable
layer governs the permeability level.   Otherwise, the permeability
level should be determined by the layer containing the water table.   A
set of rules are implemented to assign a permeability level for
stratified soil according to the facts described above.   These rules
can be summarized as:

IF   soil stratification is observed
   and there does not exist a less permeable layer above
           the soil layer containing the water table
THEN   the permeability level is determined from the soil
           layer containing the water table
       because horizontal flow governs.

IF   soil stratification is observed
   and there exists a less permeable layer above the soil
           layer containing the water table
THEN   the permeability level should be determined from
           the less permeable layer
       because vertical flow governs.

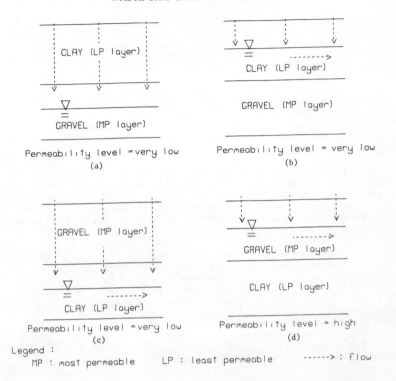

Figure 1 : Determination of Permeability Level for Two Layer System.

Although these two simple rules-of-thumb only use two layers of soil, they apply to general multilayer systems. One of the advantages of these two simple rules is that they require information pertinent to the layers of interest only. An example for the determination of permeability level using the current prototype program is shown in Appendix I.

Determination of Ground Water Flow Direction and Gradient

Ground water flow direction and gradient are basic items of information required in the hazardous waste site characterization process. Ground water flow direction is often determined by observation (especially during a Phase I investigation), rather than by experimentatioin, by studying the location of existing water bodies such as wells, springs, lakes and streams, and by studying site topography. These surface features provide valuable information that an engineer often uses during preliminary investigations.

In Phase I type investigations, surface topography and the elevations of nearby water bodies will usually be the prime source of data, since ground water monitoring wells are generally not installed. However, Phase II investigations almost always include the installa- tiion of ground water monitoring wells, and thus ground water table elevations will be available.

In general, ground water flow direction and gradient can be estimated by using three given locations with known water table elevations surrounding the site. The method is generally known as the three point method, which is an algorithmic procedure. This procedure is implemented as an external function in LISP and called by the OPS5 program as follows:

```
(p flow==set-up-call-three-point-method
    {<goal> (goal ^objective flow ^action analysis
                  ^status three-point)}
    {(vector ^attribute <x1> <x2> <x3>
             <y1> <y2> <y3> <z1> <z2> <z3>) <vector>}
-->
    (call tpoint (substr <vector> 1 inf))
    (modify <goal> ^status complete))
(defun tpoint ()
    (setq count ($litbind 'attribute))
                      :
    [...set up vectors xx, yy, zz...]
                      :
    (desort xx yy zz)              ;;descendent sort by zz
    (setq direction (sbrg xx yy zz))   ;;compute bearing
    (setq grad (sgrad xx yy zz))       ;;compute gradient
    ($reset)
    ($value 'flow_result)          ;;
    ($tab 'gradient)               ;;put results into
    ($value grad)                  ;;working memory
    ($tab 'bearing)                ;;
    ($value direction)             ;;
    ($tab 'status)                 ;;
    ($value 'true)                 ;;
    ($assert))
```

Although three locations with corresponding elevations are sufficient to estimate the ground water flow direction and gradient (three points define a plane), additional water bodies of interest can be used to check the estimate. A typical run of the current prototype system to estimate ground water flow direction and gradient is shown in Appendix II.

Discussion and Future Work

The expert system, when completely developed, will be able to characterize a site in terms of the possibility and extent of ground water contamination, surface water pollution and air pollution. The model presented in this paper is still under development, although at present it is quite adequate for HRS scoring. The complete system will consider a wider variety of factors, and will consider how these factors work as a system in ground water contamination. For example, the system will independently evaluate information about the aquifer depth, soil permeability, ground water direction and gradient, well locations and populations they serve. The system will then combine the information to estimate how long it will take for contaminated ground water to reach nearby wells and well fields, and estimate the population threatened by the contamination.

The resolution of conflicts is a very important task. Rarely does all the data obtained in a hazardous waste site investigation lead to one firm conclusion. Several interpretations are usually possible, however the expert resolves the conflicts and arrives at the most probable conclusion. Interestingly, the rules utilized at this stage are usually qualitative rather than quantitative in nature, and tend to be quite simple rather than complicated. However, in the process of developing expertise, the expert acquires a large number of these rules. The implementation of all these rules in an expert system is no simple task.

Future developments and enhancements to the prototype system will include:

(1) Consideration of the type of facility and containment methods used.

(2) Consideration of the chemical nature of the waste, including toxicity and persistence.

(3) Consideration of contaminant transport mechanisms such as dispersion and diffusion.

(4) Consideration of the quality of data. This can be accomplished by assigning a confidence value to the data and computing an overall confidence value to show the reliability of the results.

(5) Consideration of the sensitivity of the data. The model will consider how more data or better quality data will effect the final results. Often, the acqusition of additional data, at considerable expenditure of time and money, will not effect the HRS scoring. This can be important in decisions related to the allocation of investigation dollars.

(6) Explanation of the results. This would include the criteria and logic the system used to arrive at the results. This will provide documentation, and also serve as a final check on the results (experts may disagree with some conclusions).

References

1. Uncontrolled Hazardous Waste Ranking System - A User's Manual, MITRE Corporation, FR Vol. 47, No. 137, 1982.

2. Wu, Jy A and H. Hilger, "Evaluation of EPA's Hazard Ranking System," Journal of Environmental Engineering, ASCE 110(4):797-804, 1984.

3. Brownston, L., R. Farrell, E. Kant and N. Martin, Programming Expert System in OPS5, Addison-Wesley Publishing Company, Inc., 1985.

4. Winston, P.H., Artificial Intelligence, Addison-Wesley Publishing Company, Inc., 2nd Edition, 1984.

5. Forgy, C.L., OPS5 User's Manual, Report CMU-CS-81-135, Department of Computer Science, Carnegie-Mellon University, 1981.

Appendix I Example for Determination of Permeability Level

The following is a listing of a terminal session of the prototype ground water routing program written in OPS5. The example is to determine the permeability level of the two layer system shown in Figure 1(a). Input data from the user are underlined.

*(remove *)

NIL
*(make ready)

NIL
*(run)

39. START 127
****
 Ground Water Routing System
****
 Please enter objective permeability

40. PERMEABILITY==INITIALIZATION 128

DETERMINATION OF PERMEABILITY LEVEL
Please indicate if soil stratification is observed yes

41. PERMEABILITY==MAKE-STRATIFICATION-INPUT-1 130

Enter information about the layer containing the water table

42. PERMEABILITY==INPUT 132 133

Enter hydraulic conductivity if known, otherwise type nil nil

Enter soil type if known, otherwise type nil gravel

43. MAKE-CLASS==HIGH-PERMEABILITY-MATERIAL 137 135
44. PERMEABILITY==CHECK-STRATIFICATION-INPUT-2 141 139

Is there a less permeable layer above the water table? yes

45. PERMEABILITY==MAKE-STRATIFICATION-INPUT-2 141 139 142

Enter information about the less permeable layer

46. PERMEABILITY==INPUT 144 146

Enter hydraulic conductivity if known, otherwise type nil nil

Enter soil type if known, otherwise type nil clay

47. MAKE-CLASS==VERY-LOW-PERMEABILITY-MATERIAL 150 148
48. PERMEABILITY==STRATIFICATION-PRINT-RESULT-2 154 139 152

Soil stratification is observed.
Vertical flow through a less permeable layer is observed.

Given that

For the layer containing the water table
 1 Soil material is GRAVEL
 2 Hydraulic conductivity is NIL
 3 Permeability is HIGH based on SOIL-MATERIAL

For the less permeable layer above the water table
 1 Soil material is CLAY
 2 Hydraulic conductivity is NIL
 3 Permeability is VERY-LOW based on SOIL-MATERIAL

The permeability level is determined to be VERY-LOW
    and the Mitre score is 0

end -- no production true
 18 productions (106 // 276 nodes)
 48 firings (156 rhs actions)
  3 mean working memory size (4 maximum)
  1 mean conflict set size (2 maximum)
  7 mean token memory size (14 maximum)
NIL
*(bye)

Appendix II   Example for Determination of Ground Water
              Flow Direction and Gradient

Figure 2 shows an example of a hazardous waste site plan. Ground
water flow direction and gradient to be estimated from three
observation wells and a stream. A transcript of a terminal session of
the ground water routing program is listed below. Input data from the
user are underlined.

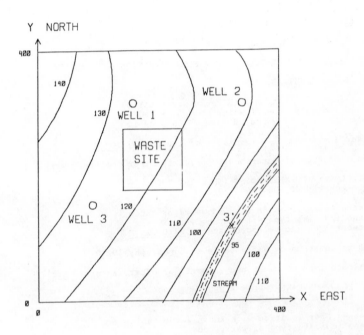

	WELL 1	WELL 2	WELL 3	STREAM POINT
X-COORD	165.0	325.0	100.0	305.0
Y-COORD	325.0	325.0	150.0	112.0
ELEV. W.T.	107.0	102.0	104.0	95.0
TOP OF GROUND	128.0	114.0	127.0	95.0

Figure 2   A Hazardous Waste Site Plan

```
*(remove *)

NIL
*(make ready)

NIL
*(run)

103. START 383
*****************************
 Ground Water Routing System
*****************************
 Please enter objective  flow

104. FLOW==INITIALIZATION 384

DETERMINATION OF GROUND WATER  FLOW DIRECTION AND GRADIENT

105. FLOW==INPUT 386 387

Please enter location with known water table
   and indicate type of water body
Enter the x- and y- coordinates  165.0 325.0

Enter elevation of water table or topography 107.0

Indicate type of water body, "well" or other features  well

106. FLOW==MAKE-NEXT-POINT-INPUT 392 390
107. FLOW==INPUT 394 396

Please enter location with known water table
   and indicate type of water body
Enter the x- and y- coordinates  325.0 325.0

Enter elevation of water table or topography 102.0

Indicate type of water body, "well" or other features  well

108. FLOW==MAKE-NEXT-POINT-INPUT 401 399
109. FLOW==INPUT 403 405

Please enter location with known water table
   and indicate type of water body
Enter the x- and y- coordinates  100.0 150.0

Enter elevation of water table or topography 104.0

Indicate type of water body, "well" or other features  well

110. FLOW==MAKE-ANALYSIS 410 408
111. FLOW==MAKE-VECTOR 412 388 397 406
112. FLOW==SET-UP-CALL-THREE-POINT-METHOD 415 413
113. FLOW==MAKE-OUTPUT 418 408 416
114. FLOW==OUTPUT 420 422

Ground water flow for set no.  1  is determined to have
Flow direction :  SOUTH-EAST
Flow gradient  :  0.042463128
```

115. FLOW==MAKE-ASK-NEW-INFORMATION 424

Do you want to replace any points? <u>yes</u>

116. FLOW==MAKE-REPLACE-POINT 428 408

How many points to be replaced? <u>1</u>

117. FLOW==MAKE-REVISE-INITIALIZATION 432 430

Enter point to be replaced <u>3</u>

118. FLOW==MAKE-REVISE-REMOVE 436 434 406
119. FLOW==INPUT 439 434

Please enter location with known water table
   and indicate type of water body
Enter the x- and y- coordinates <u>305.0</u> <u>112.0</u>

Enter elevation of water table or topography <u>95.0</u> .

Indicate type of water body, "well" or other features <u>stream</u>

120. FLOW==MAKE-REVISE-ANALYSIS 444 442
121. FLOW==MAKE-VECTOR 446 388 397 440
122. FLOW==SET-UP-CALL-THREE-POINT-METHOD 449 447
123. FLOW==MAKE-OUTPUT 452 442 450
124. FLOW==OUTPUT 454 456

Ground water flow for set no. 2 is determined to have
Flow direction : SOUTH-EAST
Flow gradient : 0.04751914

125. FLOW==MAKE-COMPARE-RESULT 458 442

Do you want to compare result with previous set? <u>yes</u>

126. FLOW==COMPARE-INITIALIZATION 462

Which set do you want to compare the current result with? <u>1</u>

127. FLOW==MAKE-COMPARISON-BEARING-MATCH 464 442 426 460

The ground water flow directions from set no. 1
   and the current set no. 2 agree.
The difference in gradient from the two sets is -11.906815 percent.

end -- no production true
 20 productions (156 // 343 nodes)
127 firings (466 rhs actions)
  7 mean working memory size (12 maximum)
  1 mean conflict set size (2 maximum)
 15 mean token memory size (35 maximum)

NIL
*(bye)

An Expert System for Flood Estimation

David Fayegh[1] and Samuel O. Russell[2], M.ASCE

Abstract

Expert systems are being increasingly applied in engineering for solving complex real-world problems. While conventional programs organize knowledge on two levels: data and program; most expert programs organize knowledge on three levels: data, knowledge base, and control. Thus, what distinguishes such a system from conventional programs is that in most expert systems the problem solving model is treated separately rather than appearing only implicitly as part of the coding of the program.

A study was made to investigate the feasibility of expert systems for solving certain engineering problems. A prototype system was developed to demonstrate the applicability of these methods to the solution of typical design flood estimation problems. It is concluded that over the long term, this type of approach to building problem-solving models of the real world results in models that are easier to develop and maintain than conventional computer programs.

Introduction

Expert computer programs have recently emerged as practical problem-solving tools. An expert system is a knowledge-based program that attempts to imitate the problem- solving behaviour of a human expert. Expert systems and other AI techniques are rapidly spreading into all professions including civil engineering. As an example of this recent popularity, in July 1985, over 10,000 people attended a conference on Artificial Intelligence (AI) in Los Angeles. At the same conference 5 years ago, attendence was less than 900. This increased interest in AI is largely due to the proven effectiveness of expert systems in solving problems in specialized domains at expert levels of performance.

Civil engineers have been using and developing conventional computer programs for many years. Many of the programs in general use are quite sophisticated and allow their users to work at a higher level of expertise than would otherwise be possible. Questions that naturally arise about expert systems include: How are they different from existing conventional programs? How can they be applied in Civil Engineering? and Would we have to discard all our existing

[1] Graduate Student, Dept. of Civil Engineering, University of British Columbia, Vancouver, B.C., Canada, V6T 1W5.
[2] Professor, Dept. of Civil Engineering, University of British Columbia, Vancouver, B.C., Canada, V6T 1W5.

programs and start recoding from scratch? To address these questions
and to investigate the usefulness of expert systems for civil
engineering applications, a pilot system was set up to solve a
typical problem in hydrology, namely that of estimating floods with
specified return periods for use in design.

The problem of estimating design floods is typical of many civil
engineering problems. Standard computational procedure are
available, many already in the form of computer programs. But
experience and judgement are usually required in selecting the most
appropriate method and values of the parameters to be used in the
models selected. In this paper, the flood estimation problem is
first described. Next the main components of expert systems are
outlined and the features that are likely to be important in civil
engineering applications are identified. Then the pilot expert
system is described and suggestions are made about future
directions.

The Problem

     For many structures such as storm-sewers, culverts, bridges, and
dams, the maximum flow that the structure has to pass is an important
parameter. Usually this flow is specified by its return period.
This is the period during which, on average, the flow will not be
exceeded. Typical values range from 2 years for very minor
structures where overtopping would be of little consequence, to 200
years for culverts on major highways. For structures such as large
dams where the consequences of failure could be catastrophic, a
bounding approach is employed and the "maximum probable flood", the
flood that "almost certainly" cannot be exceeded, is used for design.
This study focussed on estimating design floods with specified return
periods.

     There are several standard methods available for estimating
design floods but in most circumstances it is not obvious which
method is best. A procedure which is appropriate in one location may
produce poor estimates in another. The most suitable method depends
on many factors such as the importance of the project, consequences
of failure, available data and so on. Since explicit criteria have
not been defined, selection is made on the basis of the hydrologist's
judgement which, in turn, depends on his or her expertise.

     Conventional computer programs are available for most of the
standard methods used in computing design floods. These vary in
quality from non-interactive FORTRAN programs, which read the
characteristic data (such as the maximum annual floods) from some
rigidly formatted file, to "user friendly" programs in which the
information is obtained interactively before the calculations are
carried out. At the beginning of this study, 3 such programs were
already available:

1.   FDRPFFA - this FORTRAN program was originally developed by the
Water Survey of Canada and performs conventional frequency analysis
on peak annual streamflow data. Four different probability

distributions are fitted to the data series using both the method of moments and/or maximum likelihood to estimate the distribution parameters. The program also produces plots of discharge versus the recurrence interval, showing the fitted curves and the observed streamflow data.

2.   FLOODS - an interactive program (written in 'C') developed at the University of British Columbia for estimating design floods from data that would usually be considered inadequate. It allows for subjective input in the form of low, most probable and high estimates of the mean and coefficient of variation of annual peak floods; provides for updating these estimates in the light of any other available data using Bayes' theory; allows for conversion of mean daily peak flows to instantaneous peaks and generates a list of design floods with the corresponding return periods (Russell, 1982).

3.   EPFFM - an interactive program (written in FORTRAN 77) to compute flows from rainfall data, based on the Clark model. It allows for input in the form of low, probable, and high estimates of 13 user supplied parameters describing the physical characteristics of the drainage basin and provides estimates of confidence limits on the computed peak flows. It was developed as a graduate project at U.B.C. (Zachary, 1985).

The interactive programs FLOODS and EPFFM represented attempts to write programs that would not only be useful in themselves but also be easy to use. With these, options are provided in the form of menus and data is entered in response to prompts. This contrasts with the older type of program where the data entry has to conform to a specified format. Although interactive programs are easy to use, there is a price to be paid in that they require much greater effort during development. It is generally accepted that at least 50% of the effort goes into the dialogue part of the program to make it "user friendly"; but in the authors' experience, this figure has been closer to 90%. Also, it was found that the amount of effort increases rapidly with program length. Short interactive programs are fairly easy to write but long ones get progressively more difficult, mainly because of the difficulty of keeping track of all the options. Beyond a certain point, the difficulties become almost insurmountable.

The aim of this study was to check on the possibility of developing a program or system that could provide a user with advice on which approach would be best for estimating design floods given the actual circumstances at the time; be able to give advice on the running of other programs and assist its users in the selection of suitable values for the model parameters; call up and run the appropriate program; and provide help in interpreting the model output.

Expert Systems

"Expert systems" are computer programs that use specialized knowledge to solve complex real-world problems at a level of

performance comparable to that of a human expert. The main feature
that distinguishes an expert system from an ordinary program is that
the domain-specific problem-solving knowledge used by the program is
explicitly included as data in a "knowledge-base", rather than
appearing only implicitly as part of the coding of the program (Nau,
1983).

In a conventional computer program, problem-solving knowledge is
usually organized on two levels: data and procedure. A procedure is
a sequence of actions to be performed on input data in order to solve
a specific problem. When the problem solving procedure is well
understood and there is a need for repetitive use of the same
procedure, this expertise can be conveniently captured in a
conventional program. Such programs follow step-by-step instructions
to achieve their goals. However, when the precise sequence of action
necessary for solving a particular problem is not known in advance,
or when there are many potential methods that could provide a
reasonable solution, a search for the most appropriate method must be
conducted over a large number of possible solutions. Under these
circumstances, it is convenient to organize the problem-solving
knowledge on three levels: data, knowledge base and control - the
usual form of an expert system.

In expert systems, knowledge on the data level is organized in
somewhat the same way as in a conventional interactive program.
Information is transferred by the control program, from the knowledge
base, to the working memory, which is then modified as additional
data is provided by the user in response to questions. Thus at any
point during a consultation, this data represents the expert system's
knowledge about the current program under consideration. This
component of an expert system is sometimes referred to as the
"working memory", "work space" or "context".

The knowledge base is a special file which contains knowledge
organized in a variety of data structures such as "production rules",
"semantic networks", "frame networks", and other knowledge
structures. These data structures are used to represent the
domain-specific problem-solving knowledge that would be used by a
human experts in solving a particular class of problems. The
knowledge base may also reference procedures which can be called up
by the control program to solve well-defined sub-problems.

Knowledge on the control level may be thought of as a set of
general-purpose reasoning, or search procedures used to manipulate
the contents of the knowledge base to solve specific problems. The
control program, which is sometimes called an "inference engine", may
be highly centralized or distributed over a number of general-purpose
"operators". The type of control strategy depends on the contents
and organization of the knowledge base and on the domain of
application.

Expert consulting systems have been most successfully applied
for diagnosis problems such as machine fault diagnosis, disease
diagnosis and so on. In such systems, the main part of the knowledge

base consists of a set of rules in the form of "IF <Condition
1..and.. Condition j> THEN <Conclusion 1..and.. Conclusion>", rules
known as "productions". It is the responsibility of the control
program to determine which rules are applicable to which facts and
'when' and 'where' to apply a certain sequence of "operators" in
order to drive the general reasoning behaviour of the system. There
are two main inference mechanisms used with production systems,
namely forward and backward chaining. Forward reasoning (or
chaining) involves scanning through the condition part of each rule,
until one is found that matches the information provided by the user.
The objective is to move forward from the initial state to a final
conclusion such as the most probable fault in a machine. Backward
reasoning begins with the selection of a top level goal or hypothesis
to be tested. The control program then scans through the rules in
the knowledge base to find the rule whose action part can achieve the
goal. Once such a rule has been found, the program tries to confirm
the condition part of that rule by making its conditions a sub-goal
and re-scans the condition parts of the remaining rules for a match.
If no match is found an alternative hypothesis is selected for
testing. This recursive process is continued until all the
conditions are satisfied at which time the rule is applied and
hypothesis is confirmed (Duda, 1981). Modularization of knowledge
into a coherent set of rules and facts is the best method for
representing certain types of expert knowledge. These systems are
most useful for answering questions such as "What is ----" but are
not suited for answering "How do you ----" [Thompson, 1985].

The ability to add to, or change the knowledge base without
having to change the control program is an important aspect of expert
systems. Much of the early AI research was directed at building a
general problem solver which could solve a large variety of complex
problems using rules of logical inference in conjunction with fast
computers [Hayes-Roth, 1984]. However, it was realized that such
systems were not capable of solving complex real world problems
without specialized knowledge. Thus expert systems must usually be
specifically tailored to the problem area for which there is a need
for expert judgement and advice.

The key concepts for building expert systems include separating
the knowledge from the control program as much as possible and
setting it up so that the knowledge can be added to or changed with
minimal effort. However, since most engineering problem solving
involves at least some numerical computations, an engineering system
should have facilities for doing calculations as well as making
inferences. The following section outlines the pilot system FLOOD
ADVISOR developed to provide advice and assistance in estimating
design floods.

The Pilot System

FLOOD ADVISOR is a pilot expert system used to provide
interactive advice about design flood estimation. The first step
involved classifying the estimation problem into 5 generalized
categories:

CASE 1: A "relatively long" period of record is available on the
         stream of interest and the recording station is within the
         basin of interest.

CASE 2: A "relatively short" period of record is available on the
         stream of interest and the recording station is within the
         basin of interest.

CASE 3: Regional streamflow data is available for the region.

CASE 4: Precipitation data is available for the region of interest.

CASE 5: No streamflow or rainfall data are available for the region
         of interest.

The control program for FLOOD ADVISOR was initially written in
'C' by Oberski and Zittini at the University of British Columbia.
This program was adapted to run on a VAX/730 minicomputer under the
Unix (TM of Bell Laboratories) operating system as part of a graduate
project (Fayegh, 1985). The knowledge base is a file with the
following format:

```
start "!proc0" > node1 node2 node3 <
node1 "!proc1" > node1.1 node1.2 node1.3 <
node1.1 "!proc1.1" > node1.1.1 <
node1.1.1 "!proc1.1.1" > <
node1.2 "!proc1.2" > <
node1.3 "!proc1.3" > node1.3.1 <
node1.3.1 "!proc1.3.1" > <
node2 "!proc2" > node2.1 node2.2 <
node2.1 "!proc2.1" > node2.1.1 <
node2.1.1 "!proc2.1.1" > <
node2.2.2 "!proc2.2.2" > <
node3 "!proc3" > node3.1 node3.2 <
node3.1 "!proc3.1" > <
node3.2 "!proc3.2" > node3.2.1 <
node3.2.1 "!proc3.2.1" > <
```

The program reads the above file as initial input and sets up an
open-ended tree structure. Users start with a main menu that lists
the major options. Once an option is selected, the control program
causes the execution of a procedure attached to that option. To
further illustrate, consider line 1 in the above knowledge base:
start "!proc0" > node1 node2 node3 <. The procedure attached to the
root node is enclosed in quotation marks (i.e. "!proc0") and is
invoked when the user enters 'start'. This attached procedure may be
either an operating system command calling upon Unix facilities (such
as the editor), external programs, or a short prompt. The remainder
of the line which is enclosed in brackets (i.e. > node1 node2 node3
<) represent the options available to the user at the root node.
Upon entering a valid option (say node3) the program transfers
control to the named node and causes the execution of the
corresponding procedure (i.e. "!proc3"). This process is repeated
until a goal node has been reached.

As mentioned previously, the primary purpose of the FLOOD ADVISOR is to provide interactive advice about the flow estimation technique most suitable to one of the above five situations. For example, suppose that it has been established that a case 1 situation is applicable to the problem at hand. The following tasks are then performed:

(1) The user is informed about the assumptions made about the nature of streamflow data (such as statistical independence, homogeneity, etc.) and the appropriate procedures are invoked to verify each assumption.

(2) Once it has been established that all assumptions are satisfied, the user is advised to use conventional frequency analysis with the FDRPFFA program. Upon request, the user is provided with simple instructions on how to input his/her data according to the appropriate format. To avoid entry error, a sample input file is invoked on the editor and the user is asked to input his data in this file by simply "writing-over" old entries. After all streamflow data has been entered, FDRPFFA is automatically run and the user has the option of either obtaining a hardcopy of the output and the generated plots or viewing the result on the screen.

(3) The last phase of the consultation consists of providing advice on how to interpret the output from the previous step.

FLOOD ADVISOR treats the other 4 cases in a similar manner - by providing advice and calling up the most suitable application program.

Summary and Conclusions

A pilot expert system was developed for estimating design floods. The flood estimation problem was broken down into five generalized situations and a program was set up to identify the appropriate category and recommend a solution procedure depending primarily on the type, quality and quantity of the available data. The system then provides advice on how to use the recommended procedure, calls up and runs the appropriate application program and finally provides advice on interpreting the output.

The system consists of a knowledge base and a control program. The knowledge base is encoded in a file that can be easily modified. The control program "walks through" an open-ended tree, used to represent the domain specific knowledge, and can also call up and run self-contained applications programs, written in any other programming language supported by the operating system. The control program can be adapted to incorporate both forward and backward chaining as attached procedures. Also a copy of the control program, with its own specialized knowledge base, could be used as an attached procedure. This feature would allow nesting of sub-trees and facilitate modular system development. This approach thus offers considerable versatility and the opportunity to incorporate a large amount of problem solving knowledge within a single system.

At present, knowledge is added to the knowledge base by direct insertion in the file. An interactive graphics procedure is planned to facilitate visualization of the tree structure during knowledge acquisition. In addition, the system cannot directly incorporate the results of numerical computations into its line of reasoning, but it is planned to add this facility to make it into a more useful general purpose tool for engineering applications.

In summary, the expert systems approach simply represents a more efficient way of organizing available knowledge in the form of computer programs and data structures for solving complex problems than conventional programs offer. In this sense, they could be considered as a step beyond the structured programming concepts that noticeably advanced the art of programming. But it is a giant step, that promises to greatly reduce programming costs and ocmplexity and seems certain to have a large influence on civil engineering.

References

Duda, R. & Gashing, J.G. (1981). "Knowledge-Based Expert Systems Come of Age", Byte, Vol. 6, #9, Sept. 1981, pp. 238-279.

Fayegh, A.D. (1985). "Flood Advisor: An Expert System for Flood Estimation", M.A.Sc. Thesis, Dept. of Civil Engineering, U.B.C.

Hayes-Roth, F., Waterman, D. & Lenat, D. (1984). "Building Expert Systems", Addison-Wesley Co.

Nau, D.S. (1983). "Expert Computer Systems", Computer, Vol. 16, February 1983, pp. 63-85.

Russell, S.O. (1982). "Flood Probability Estimation", Journal of the Hydraulics Division, Proc. of ASCE, Vol. 108, #HY1, Jan. 1982, pp. 63-73.

Thompson, B. and Thompson, W.T. (1985). "Micro Expert", McGraw Hill, New York.

Zachary, A.G. (1985). "The Estimated Parameter Flood Forecasting Model", M.A.Sc. Thesis, Dept. of Civil Engineering, U.B.C.

An Expert System for Site Selection

Irene T. Findikaki[*]

Introduction

Site selection is a critical element in decision making for a wide range of economic activities, from land use planning to residential location selection by individuals. Typically, site selection involves two phases. First, a small number of alternatives among all possible sites are identified for further evaluation, and then the selected alternatives are examined in depth to find the optimal site. Because of the effort required for a thorough evaluation of each site, the number of sites included in the second stage is, by necessity, small. Therefore, the first phase of the site selection process is very important, because if, for example, the "best" available site is not included in the set of alternatives selected for thorough study, then this site can never be selected as the optimal site. In large scale planning the first phase involves reconnaissance and prefeasibility studies. At the individual's level it may involve consultation with a specialist.

This paper presents a methodology for systematizing and expediting the first of the above two phases in the site selection process. The proposed methodology is implemented in the expert system SISES (SIte Selection Expert System) that can be used as a decicion making tool in selecting a site for a specific use. SISES is designed to capture the expertise of expert decision makers without making any a priori assumptions about their preferences, background or identity. For example, the decision maker can be a land use planner, a developer or a prospective land user. Although the system was developed to reproduce a specific domain of expertise, it can easily be adapted to other applications with a

[*] RFT Associates, PO Box 3011, Stanford, California 94305-0029.

similar conceptual framework independently of domain. The system has the ability to analyze and learn about the decision making process of individuals.

SISES is written in Pascal and runs on a microcomputer. The present paper gives an overview of the ongoing development of SISES, and focuses on the knowledge acquisition methodology.

Structure of SISES

SISES consist of four parts, the knowledge acquisition, design, induction, and decision analysis units. The knowledge acquisition unit is used to collect and organize information provided by expert decision makers. The induction unit evaluates this information and generates rules and entity evaluation functions expressing the judgement of the experts. Knowledge acquisition in SISES is based to a great extent on a multiple trade-off data gathering procedure and requires a close interaction between the knowledge acquisition and the induction units. The design unit offers the capability to customize the knowledge acquisition system to better match the needs of particular applications by modifying selected elements of the knowledge acquisition unit, and expanding it with the addition of new components. The decision unit uses the rules generated by the induction module and employs decision theory techniques for selecting one or more of the available alternatives. The overall structure of SISES is illustrated in Figure 1.

The knowledge acquired from an expert can be stored for future use, contributing thus to incremental learning. The system offers the option to combine knowledge from several experts and store it in a dynamic knowledge base which can grow as new decision makers are interviewed. The system can use either the selection rules of a single decision maker, or rules representing the idiosyncratic decision processes of a group of individual decision makers. Therefore, points of departure among the views of human experts who disagree can also be represented and taken into account.

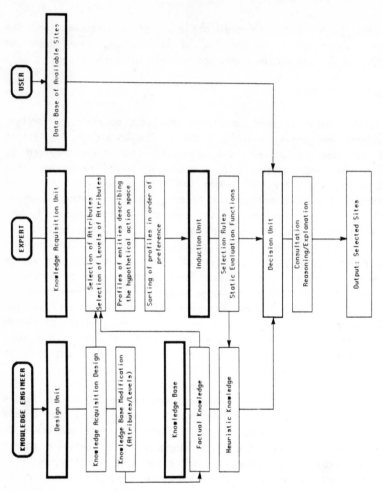

Figure 1. Schematic Structure of SISES.

The Knowledge Acquisition Methodology

Knowledge acquisition is the most important of the central problems in artificial intelligence research (Fiegenbaum, 1983). Quinlan (1979) points out that one of the reasons that knowledge acquisition is the bottleneck of expert system building is the fact that typically experts are asked to perform tasks that they do not ordinarily do, such as describing everything they knows about their domain of expertise. A key research objective in knowledge acquisition is the development of methods of machine learning to replace the conventional approach of knowledge acquisition via dialog, in which the expert authors rules during long interviews with a knowledge engineer. Machine learning involves the development of rules by analyzing cases of correct decision making, and generalizing from specific examples provided by the expert. Induction of rules by generalization from examples of expert decision making has been used successfully in some expert systems (Mitchie, et al, 1984).

The knowledge base of SISES consists of knowledge representing basic facts about site selection built in the system, and heuristic knowledge which is induced by observing and analyzing specific examples of judgemental behavior of the expert. The example cases that the expert is called to express judgement on are generated by the system in collaboration with the expert. The knowledge acquisition unit is designed to interact with the expert to acquire information relevant to both the knowledge domain and the judgemental behavior of the expert.

The knowledge acquisition unit is complimented by a design unit which can be used to modify the default data base of factual knowledge used by the system, update information about the domain, and alter the design of the knowledge acquisition unit.

The knowledge acquisition methodology is based on the following assumptions on the behavior of expert decision makers:

    a. An expert's judgement on the selection or rejection of a site for a particular use is determined by a limited number of basic factors or attributes which are a subset of all site characteristics. These factors form the profile of a site.

    b. An expert's judgement is based on the evaluation of several

alternative sites described by the levels of their attributes.
The decision process leading to the selection of a particular
site is governed by simultaneous trade-offs between pairs of
attributes and between entire site profiles. The final
judgement of the expert is the product of the evaluation of
complete profiles of the available sites.

c. The selection of a site is made under conditions of certainty
   and it is of extreme importance to the decision maker.
   Consequently, the individual decision maker always tries to
   select the best available alternative, i.e. the alternative
   perceived as having the maximum value. This eliminates the
   need for probabilistic modeling of the expert's judgement
   (Shoker and Srinivasan, 1979; Johnson 1974). This assumption
   implies zero probability of selecting any alternative other
   than the one representing the expert's top choice. A
   consequence of this assumption is that preferences can be
   measured on an ordinal scale.

d. The perceived value of a site is a function of the
   characteristics of the site weighted by their relative
   importance to the decision maker.

e. The order of preference among multi-attribute entities
   represents information about an individual's decision process
   in condensed form. Decomposition of the overall judgement of
   an individual into its elements provides the ground for making
   inferences about the relative importance of each attribute and
   for analyzing the trade-offs which characterize the decision
   process.

f. Although direct statements by decision makers about
   preferences, rules of thumb, and comparison procedures are
   valuable, in most cases, it is uncertain whether the decision
   makers are fully aware of the facts, prejudices, beliefs and
   heuristic knowledge influencing their judgement. Declarative
   statements do not always provide the information needed to
   reproduce the decision maker's behavior.

The description of the judgemental process of decision makers is
achieved in SISES by combining information based on declarative
statements and information induced from observations of the
behavior of decision makers working on specific examples. The
knowledge acquisition unit is designed to provide the system with
both types of information. The latter type of information is
obtained with the use of a special method designed to reveal the

preference structure of individual decision makers. Some of the above assumptions on the cognitive process are incorporated in a function which expresses the outcome of the judgemental evaluation of individual entities within a hypothetical action space. This function, which quantifies the value of individual entities to the decision maker, is termed the entity evaluation function.

The hypothetical action space is an abstraction of a subset of the real action space of the decision maker. The real action space is defined as the part of the environment that the individual has contact with (Wolpert, 1965). The concept of the action space is similar to the concept of "life space" introduced by Lewin (1951). The hypothetical action space consists of a finite number of entities, typically much smaller than that in the real action space, described by combinations of discrete levels of attributes representative of the real action space of the decision maker. The hypothetical test space is generated by the system and is designed in such a way as to include entities covering the entire range of attributes of the entities found in the real action space of the decision maker. For example, in the case of residential selection a hypothetical action space may consist of several profiles of residential sites, each described by attributes that are important to the decision maker, e.g price, air quality, noise level, quality of schools, density, proximity to main arteries, distance from work, etc (Findikaki, 1981, 1982). SISES uses an experimental design scheme to reduce the dimensionality of the hypothetical action space to a manageable number of dimensions. The default experimental design is a fractional factorial design (Addelman, 1962a,b).

The system provides the option of selecting the level of the knowledge acquisition session, depending on the desired level of sophistication, and time limitations and availability of the expert. In developing a sophisticated system and working with an expert whose availability is not a limiting factor a session based on the full-fledged knowledge acquisition design should be selected, while for a less sophisticated system a shorter session based on a less elaborate design could be used.

The first step in a typical knowledge acquisition session is the selection of the attributes that describe the entities forming the hypothetical action space. For this purpose, the expert is asked

first to specify the type of land use in the problem under
consideration. Then the most important among all potential site
attributes, relevant to the desired land use, are identified
through a series of questions to the decision maker. The list of
potential attributes for different land use types is part of the
factual knowledge base of the system. This list can be expanded
with additional attributes suggested by the expert. New attributes
proposed by the expert during a particular session can be used to
expand the permanent data base of potential attributes. After the
relevant site attributes are identified, the expert is asked to
distinct between quantifiable and non-quantifiable attributes and
criteria.

Next, the system asks the expert to verify the range of
potential values of the quantifiable factors. The default values
for the range of the attributes provided by the system   are
displayed and the expert is asked to define the lowest and the
highest expected level of each of the new attributes. The expert
has also the option to change any default attribute level ranges.

Then, the expert is asked to group the quantifiable attributes
in groups of six or five, and rank the groups in order of
importance. For example, in the case of a total of sixteen
attributes the expert would identify three groups of attributes,
the six most important, the five next most important, and the five
least important attributes. The attributes in each group are used
then by the system to construct profiles of individual entities
forming a hypothetical action space. Each profile is described by
the levels of the attributes. The level of each attribute is
determined with the aid of a fractional factorial design.

Then the expert is presented with the set of site profiles
defining the hypothetical action space formed by the first group of
attributes and is asked to rank these profiles in order of
preference. At this point, the system passes the ranking of the
profiles to the induction unit which, using the method described in
the next section, identifies the most important among the
attributes in this group. The most important attribute is used
together with the next group of attributes to design a second set
of hypothetical profiles which the expert is again asked to rank.
Then, the most important attribute of the first group is used with
the remaining attributes to form a third set of hypothetical

profiles and the expert is asked again to rank them.

A similar approach is followed with the non-quantifiable attributes. The efect of these attributes is expressed in the form of yes/no conditions. The expert is asked first to identify the non-quantifiable attributes and the corresponding yes/no condition that would result to the rejection of a site.

Induction Unit

The effect of the non-quantifiable attributes on the judgement of the expert can be expressed by a set of rules, either based on declarative statements, or induced from specific examples of expert decision making.

The combined effect of the quantifiable attributes is described by a linear function of weighted attribute levels. This formulation is based on the assumption that the cognitive process of an individual making choices can be simulated by a linear model (Fishbein,1954; Rosenberg,1976). Thus, the value of a particular site to the decision maker is expressed as:

$$(\text{Site Value})_j = \sum_{p=1}^{n} w_p x_{jp}$$

where n is the number of site attributes, x_{jp} is the level of attribute p for site j, and w_p is the importance of attribute p. The value of a site may increase as the level of each attribute p increases (if $w_p > 0$), or as the level of attribute p decreases (if $w_p < 0$), or it may be independent of the level of p (if $w_p = 0$). The selection criterion is that the expert selects the site which has the largest value of the evaluation function

The order of the profiles constitutes a set of conjoint measurements reflecting the joint effect of different attributes on judgements (Tversky, 1967a,b). The order of preference of the profiles is converted to paired judgement comparisons which are analyzed to obtain the importance weights of the attributes forming the entity evaluation function of the individual. The importance weights are estimated in such a way as to reproduce the order given by the expert. This is achieved by using a linear programming formulation (Shoker and Srinivasan, 1977) in which the objective is to minimize the error in reproducing the expert's ordering of the profiles and the constraints are based on the paired judgement comparisons.

This method is applied to all sets of ordered profiles. Then, the most important attribute which is included in the description of all hypothetical spaces is used as a bridging factor to combine the evaluation function for each action space into a global evaluation function.

Decision Making Unit

The decision making unit evokes different scripts matching the needs of the user. These scripts are composed by the system using a combination of built in information, and the induced rules and entity evaluation functions.

During consultation the user is asked to provide a description of all potential sites in terms of the attributes used by the system for the particular land use under consideration. The system first makes inferences to reduce the data base of potential sites by eliminating those that should be rejected according to the induced rules based on the non-quantifiable site attributes. Then, the system computes the entity evaluation function for each of the remaining sites. The output of the consultation session consists of the order of preference of the selected sites and an optional explanation for rejecting the rest.

Summary and Conclusions

SISES is an expert system for site selection that can be used as a decision tool in land use planning, residential selection and a other site dependent planning. The system consists of four units for knowledge acquisition, rule and evaluation function induction, decision making, and design of customized versions of the knowledge acquisition unit. The emphasis of the present work is on the description of the knowledge acquisition unit. The knowledge acquisition methodology used by SISES is based on the interactive generation of a set of profiles describing a hypothetical action space that the expert is asked to rank. The ranking of these profiles is decomposed to paired comparison judgements which constitute a set of examples of expert decision making. Analysis of these examples leads to the induction of site selection rules and the generation of an entity evaluation function. The selection rules and the entity evaluation functions are applied on the data base with the chaaracteristics of all available sites provided by the user to generate the list of selected sites, ranked in order of preference.

References

Addelman, S. 1962a: "Orthogonal Main-Effect Plans for Asymmetrical Factorial Experiments", Technometrics, vol 4, pp 21-26.

Addelman, S. 1962b: "Symmetrical and Asymmetrical Fractional Factorial Plans", Technometrics, vol 4, pp 47-59.

Fiegenbaum, E.A., 1983: "Knowledge Engineering: The Applied Side", in Intelligent Systems, J.E. Hayes and D. Michie, eds., Ellis Horwood Limited, Chichester, England.

Findikaki, I.T., 1980: Residential Location Preferences: A Conjoint Analysis Approach, Engineer Thesis, Stanford University.

Findikaki, I.T., 1982: Conjoint Analysis of Residential Preferences, PhD Dissertation, Stanford University.

Fishbein, M., 1967: "A Behavior Theory Approach to the Relations between Beliefs about an Object and the Attitude towards the Object", in Readings in Attidue and Theory Measurement, M. Fishbein ed., John Wiley and Sons, New York.

Johnson, Richard M. (1974) "Trade Off Analysis of Consumer Values" in Journal of Marketing Research, vol XI May 1974.

Lewin. K., 1951: "Field Theory in Social Sciences", Harper and Row, New York.

Michie, D., S. Muggleton, C.Riese and S. Zubrick, 1984: "Rulemaster: A Second Generation Knowledge-Engineering Facility",

First Conference of Artificial Intellegence Applications, Denver, December 5-7.

Quinlan, J.R., 1979: "Discovering Rule by Induction from Large Collections of Examples" in Expert Systems in the Micro Electronic Age, D.Michie ed., Edinburgh University Press.

Rosenberg, M.J., 1956: "Cognitive Structure and Attidunal Affect", Journal of Abnormal and Social Psychology, vol 53, pp 367-372.

Tversky, A., 1976: "A General Theory of Polynomial Conjoint Measurements", Journal of Mathematical Psychology, vol 4, pp 1-20.

Tversky, A., 1976: "Additivity Utility and Subjective Probability", Journal of Mathematical Psychology, 4, pp 175-201.

Wolpert, J. 1965: "Behavioral Aspects of the Decision to Migrate", in Papers of Regional Science Association, vol 15, pp 159-169.

Developments in Expert Systems
for Design Synthesis

John S. Gero, Member and Richard D. Coyne*

The applicability of expert systems to design synthesis is demonstrated. This is
achieved by means of inference rules which can interpret design specifications in
order to produce designs. This approach is applicable to certain classes of design
problem which can be subdivided into independent subproblems. An expert system
which contains knowledge about conflict resolution applicable to the more general
class of design problem is demonstrated.

Experts systems are considered here as automated reasoning systems which attempt to
mimic the performance of the reasoning expert. Because of the shortcomings of formal
logic as a means of modelling human intelligence, the reasoning systems developed often
incorporate 'enhanced logics'.

Workable systems can be devised which operate on the basis of formal reasoning,
however. This is particularly so in the case of interpreting the properties and
performances of buildings where the theory by which interpretations can be made is well
understood. This is generally the case, for example, when evaluating the performance of
buildings for compliance with the requirements of building codes.

Expert systems of this type are also applicable to the synthesis of designs,
particularly for those classes of design problem which can be subdivided into independent
subproblems. But expert systems which are applicable to the more general class of design
problem can also be devised. The paper describes developments in expert systems for
design synthesis by drawing on the work of the Computer Applications Research Unit in
the University of Sydney.

The Interpretation of Designs

The facts which describe an object, such as a building, and the knowledge by which
properties of the object can be derived can be modelled as formal axiomatic systems. The

* Professor and Research Student, respectively, Computer Applications Research Unit, Department of
Architectural Science, University of Sydney, Sydney, NSW 2006, Australia.

advantage is that knowledge becomes amenable to formal proof procedures and the mechanism of logical inference (Kowalski, 1983).

A design possesses attributes other than those facts by which it is represented. These attributes can be described as derived, or implicit attributes, and a set of such attributes constitutes the semantic content of the design. The major operation in discovering meaning is interpretation by *inference*, and the knowledge about interpretation can be formalised as inference rules.

Inferential knowledge can be embedded within algorithms or represented in a form which makes the mappings between description and semantic content explicit. A useful and intuitively appealing method of representing inferential knowledge is as rules of the form:

$$\text{IF } a_1, \dots a_n \text{ THEN } b_1, \dots b_n.$$

This states that if a design possesses attributes $a_1, \dots a_n$, (the antecedents) then infer that attributes $b_1, \dots b_n$ (the consequents) are also true. The knowledge by which a design is interpreted may consist of many such rules. The attributes are propositions, and a tree, or network, of interconnected propositions is produced. Expert systems can be created which facilitate the representation of inferential knowledge in this form and make it amenable to automation.

For any system the issue arises as to how the process of interpretation should proceed. For a realistic set of inference rules the number of facts that can be derived is likely to be very large. There will therefore be more work involved in asking of a design: 'What attributes can be inferred from its description?'; than asking: 'Does the design have this particular set of characteristics?'. The former suggests a data-driven approach which starts with a design description and an attempt is made to infer as much as the rules will allow. The latter is a goal-directed approach which begins with various attributes and tries to discover if the design possesses those attributes.

A system containing inference rules is of value even when there are no facts constituting a building description. The 'leaf nodes' of an inference tree correspond to requests for facts about the building and so can be handled interactively by means of prompts. They could also be regarded as entry points to other axiomatic subsystems which interpret computer databases, or other building code systems, for example. When incorporated into a general purpose inference system a dialogue is produced. The derived facts constitute the meaning of the total system.

Such systems are therefore generally intended to be consultative, and ideal systems would communicate by means of natural language. Communication also concerns the issue of deciding at which semantic level the consultation should take place. If the system is to communicate with a novice, who knows little of the domain of expertise represented by the system, then there may be justification in making the leaf nodes the propositions at which questions are asked.

For experts, questions about propositions higher up in the network may obviate unnecessary search. In a session with the system the response 'how' provides an instruction to the inference system that it is to proceed to a lower descriptive level of inquiry. An explanation facility responds to the response 'why' which retraces the lines of reasoning so far employed in order to offer an explanation of why a particular question is being asked.

These control features have been implemented in an expert system shell called BUILD developed by Rosenman (1985). Knowledge about a domain is stored in the form of inference rules and inferences are propagated in an interactive manner. The utility of this system in the interpretation of designs has been demonstrated with reference to building codes (Rosenman and Gero, 1985). Certain attributes of the building, such as its fire resistance rating can be inferred as well as its compliance or non-compliance with certain sections of the code. The clauses of a building code (assuming they are consistent) and the description of a building can be modelled as an axiomatic system, the meanings of which are those collections of statements which determine its compliance. Inference systems are therefore readily applicable to the *interpretation* of designs.

Interpretation of Specifications

The question arises as to the applicability of inference systems to the *generation* of designs. Design descriptions can be interpreted by means of inference in order to derive certain properties of designs. Can the interpretation of design *specifications* produce design *descriptions*?

There are two conditions under which the utility of direct inference to design sythesis can be demonstrated. One involves the selection of design options. The second applies to designs which can be hierarchically decomposed into independent subproblems.

Retaining Wall Design

An expert system for the design of retaining walls has been developed by Hutchinson (1985), based on the BUILD expert system shell. The system contains knowledge appropriate to selecting between classes of retaining wall types and between different prototypical cross sectional designs for retaining walls. This knowledge is not formally available in reference material so it was extracted from specialist engineers in the field by means of a survey, and from the researcher's own expertise.

The system reduces the range of options available by judicious inquiry about soil and topographical conditions as well as designer preferences. When it has made a decision about the retaining wall prototype it proceeds to determine dimensions and other parametric conditions. The prototypical design is therefore refined incrementally, depending on the design specifications.

Various options by the system are presented graphically from a stored set of images associated with various propositions. The final design is also presented by means of a C program which interprets the design description graphically. The system has been implemented on a SUN Microsystems SUN 2 workstation in Quintus Prolog. Figure 1 shows a screen display of part of the dialogue where the user provides answers based on the semantics of the diagrams.

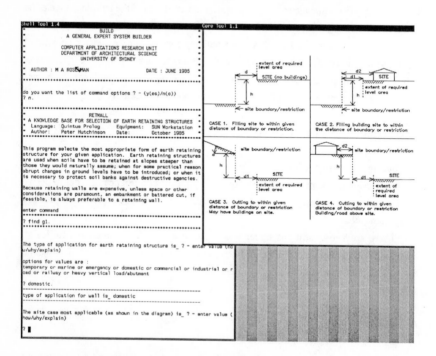

Figure 1. Screen display of part of a dialogue with an
expert system for selecting and sizing retaining walls

Figure 2 shows a screen display of the final design drawn at the commencement of the working drawing stage after the class of wall has been chosen, its cross-section selected and then finally sized for the imposed loads.

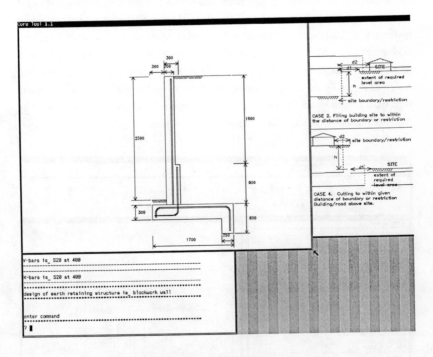

Figure 2. Screen display of the final design of the retaining
wall selected, sized and drawn by an expert system

The system demonstrates the use of a pre-determined hierarchy of problems and subproblems which are treated independently. Once the prototype has been selected, then its various subcomponents can be designed. This is a powerful approach within the substantial class of hierarchically divisible design problems, and one that is amenable to representation in automated inference systems.

Kitchen Design

A similar approach has been demonstrated in a prototypical kitchen design system developed by Oxman (Rosenman, Gero and Oxman, 1985). The intention is, however, to create three modes of interaction between a designer and the system. The first is where the designer draws a kitchen employing the modelling facility of the system and the expert system checks the design for compliance with rules of good practice. Here the model behind the graphics needs to be interpreted semantically in order to be interrogated by the expert system. This raises important questions about the interfaces between expert systems and 'traditional' computer-aided design systems. The system has been implemented on a SUN Microsystems SUN 2 workstation in Quintus Prolog. Figure 3 shows a screen display of part of the dialogue - the right-hand window shows the partial design as drawn by the designer whilst the left-hand window shows the expert system shell BUILD with its kitchen knowledge base being used to check this partial design by interrogating the model and, where necessary, the designer.

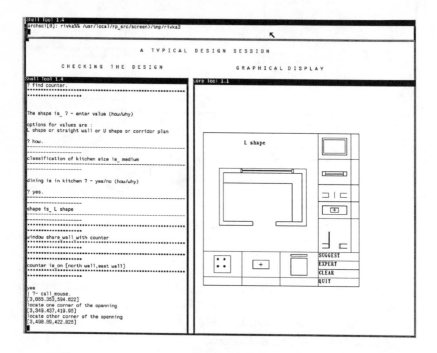

Figure 3 Screen display of an expert system checking the preliminary design of a kitchen
The input of the design is graphical

In the second mode, the expert system can check the design as it progresses and complete any detail. In the third mode the expert system makes decisions on the basis of a 'dialogue' with the designer and employs its rules of good practice to invoke and refine a prototypical solution. In this case the system first determines the appropriate shape of the kitchen and then places various components. The same knowledge is employed to check the design as that used in the selection and refinement of the prototype. Figure 4 shows a screen display of the expert system being used to produce a partial design. The left-hand window contains the dialogue between the expert system and the designer. The right-hand window shows the graphic interpretation of the selected design. The same expert system shell, knowledge base and graphics semantics interpreter is used for all three modes of interaction between the designer and the system.

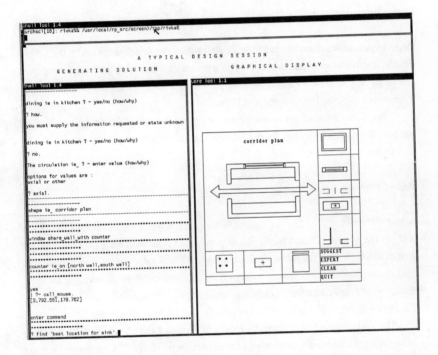

Figure 4. Screen display of the same expert system as
in Figure 3 being used to synthesise a kitchen design

In these two examples of design synthesis using automated inference the inference engine makes use of both backward and forward chaining over the knowledge base of rules. Automated inference is appropriate to design where only a relatively small number of choices are to be made and where the ranges of options are small. When there are many decisions (which is generally the case) then the interactions between decisions can cause conflicts. Such conflicts generally require the incorporation of further specialised knowledge concerning its resolution. A number of ways are being explored in an attempt to add conflict resolvers to automated inference systems. Some of these are modelled on the non-serial characteristics of the network of propositions or rules. Others add a blackboard to the system on to which constraints produced by decisions are posted so that these constraints can be propagated during chaining. A specialised resolver is employed when decisions produce constraints which are in conflict.

From Interpretations to Designs

Design systems therefore become more complex than the model proposed by the simple manipulation of propositions within inference networks. More complex models have been proposed which incorporate notions of *planning* and the exploitation of multiple abstractions (Coyne, 1985; Coyne and Gero, 1985).

Design specifications are treated here as *interpretations* for which design *descriptions* must be found. The mappings for this process are more complex than those provided by direct rules of logical inference.

Room Layout Example

A knowledge-based system for spatial layout design has been developed by Coyne (1986). The system is based on the use of production rules which are generative rather than of the type used for inference. The rules transform sets of procedures (plans) for constructing layouts rather than operating on the spaces themselves. Knowledge about the resolution of conflicts between competing actions is also represented in the form of rules. The implementation of these rules is controlled by scheduling knowledge.

The system therefore operates by means of explicit hierarchical control knowledge and exploits the inherent redundancy in certain design problems. The layout problem can be visualised as a problem of organizing nodes on an adjacency network, as sequences of actions about the placement of spaces and as the configuration of geometrical entities. Each of these views contributes to the solution, and each has its own pertinent body of knowledge. The system has been implemented on a SUN Microsystems SUN 2 workstation in Quintus Prolog. Figure 5 is a screen display of the system showing the various stages in the synthesis process. The top left window shows the tasks being processed, the bottom left window shows a network representation of the problem structure, the bottom right window shows the plan of actions and the top right window shows the resulting layout after executing the plan of actions.

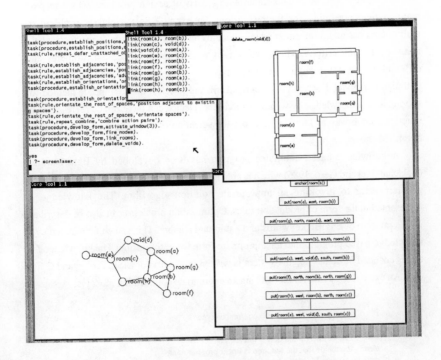

Figure 5. Screen display of the various stages in an
expert planning system for the design of room layouts

This approach resembles that of the blackboard paradigm of the HEARSAY speech understanding system (Erman et al., 1981). It appears to provide the greatest scope for the representation and manipulation of design knowledge, as it permits the representation of diverse types of knowledge. Design is a loosely structured, knowledge-rich activity and as such requires a medium for representation which allows free rein to that knowlege.

Formulation of the Decision Process

With the introduction of computerized reasoning, the problem shifts to that of structuring the knowledge into a form (a logical system) which is amenable to solution finding methods. The automated formulation of knowledge-based systems does not yet appear to be a feasible proposition in design systems such as that discussed above.

It can, however, be demonstrated with a class of design problems which rely on mathematical reasoning for their solution. An example is the classic problem of multicriteria design optimization which concerns the production of designs which satisfy many conflicting criteria. There are various models for solving such problems and different models are suitable for different types of problems. The need is to identify an optimization model described in algebraic form and to select the appropriate method for solving it. The knowledge by which methods are selected can therefore be represented in much the same way as knowledge about the selection of retaining wall prototypes as described above.

A knowledge-based computer program has been developed by Balachandran (Balachandran and Gero, 1986) which is able to recognize the structure of an optimization problem prior to invoking an appropriate solution algorithm. The knowledge is represented in the form of production rules. Optimization problems can also be described with a mixture of English keywords and a minimal algebra. This is then formulated by a knowledge-based program into an appropriate form for optimization. The knowledge for this is encoded in frames and the system is written in a hybrid of LISP, Prolog and C.

An example of a rule for recognising an appropriate optimization model is:

if	all the variables are of a continuous type
and	all the constraints are linear
and	the objective function is linear
then	conclude that the function is linear programming
and	execute LP-algorithm.

Inference rules have also been devised which recognize variables, their character, and their relationships with the constraints and the objective function.

The principle being exercised here is that of making explicit the knowledge by which an appropriate, specialised problem solving approach can be selected from a range of models, and controlling its operation by means of qualitative knowledge of the type which a designer skilled in the method might employ.

Summary

Automated inference systems are readily applicable to the interpretation of design descriptions. The question has been addressed as to whether design specifications can be similarly interpreted to produce artifacts. This appears to be the case in that class of design problems where non-conflicting decisions are to be made.

The generation of designs generally requires further knowledge than that which facilitates inferences, however. This knowledge can be afforded by generative rules and rules which resolve conflicts. Expert systems can be developed which incorporate this knowledge, though they require more complex control regimes than those afforded by standard inference systems. A direction for expert systems which reason about problem solving strategies for design has also been demonstrated. This direction leads to questions related to:

(i) how non-monotonicity in the design decision making process can be modelled using monotonic logics;

(ii) how separate knowledge bases interact at the control level;

(iii) how expert systems can interface with traditional computer-aided design systems; and

(iv) how abduction can be incorporated into a design synthesis system.

Acknowledgements

Thanks are due to those former and present members of the Computer Applications Research Unit whose work appears here: Bala Balachandran, Peter Hutchinson, Rivka Oxman and Mike Rosenman. Research support is from the Australian Research Grants Scheme, the Australian Computer Research Board, the National Energy Demonstration and Development Program and the Sydney University Postgraduate Research Studentship Scheme.

References

Balachandran, M. and Gero, J.S. (1986). Formulating and recognising engineering optimization problems. *Proc. 10th ACSMSM*, Adelaide, Australia (to appear).

Coyne, R.D. (1985). Knowledge-based planning systems and design: a review, *Architectural Science Review*, vol. 28, no. 4, pp. 95-103.

Coyne, R.D. (1986). *A Logical Model of Design Synthesis*, unpublished PhD thesis, University of Sydney, Sydney.

Coyne, R.D. and Gero, J.S. (1986). Semantics and the organisation of knowledge in design, *Design Computing* (to appear).

Erman, L.D., London, P.E. and Fickas, S.F. (1981). The design and an example use of Hearsay-III, *Proc. 7th IJCAI*, Vancouver, pp. 409-415.

Hutchinson, P. (1985). *An Expert System for the Selection of Earth Retaining Structures*, unpublished MBldgSc thesis, University of Sydney, Sydney, Australia.

Kowalski, R. (1983). *Logic For Problem Solving*, Elsevier-North Holland, Amsterdam.

Rosenman, M.A. (1985). *BUILD User Manual*, Comuter Applications Research Unit, Department of Architectural Science, University of Sydney, Sydney.

Rosenman M.A., Gero, J.S. and Oxman, R. (1985). An expert system for design codes and design rules, *Working Paper*, Computer Applications Research Unit, Department of Architectural Science, University of Sydney, Sydney.

KNOWLEDGE ENGINEERING
IN OBJECT AND SPACE MODELING

V. Tuncer Akiner*

ABSTRACT

Object and space modeling in CAD is an important issue as most
engineering and architectural design activities call for modeling of
some kind. An object or space is generally an assembly of
interdependent components. Therefore in any object or space description
and manipulation in CAD, it is of critical importance for the system to
"know" about the constituent parts of the model involved. Facts and
knowledge about the model, the attributes of its components and the
relationships that exist between the components are essential in a
knowledge based CAD system. This in turn contributes to the development
of an "understanding" on the part of the CAD system for the context in
which it is being used.

This paper gives a description of such a system, Topology-1, which
attempts to provide a tool for object and space modeling through the use
of knowledge engineering techniques. It elaborates the system
configuration, and the knowledge about topology, geometry and attributes
of objects and spaces. Further the operation of the system is described.
Finally an example is presented which demonstrates its reasoning
capability in the design appraisal of a building.

INTRODUCTION

The majority of artifacts that are designed are simply an assembly
of individual objects that are interrelated to one another. The style,
the character and the performance that is expected from a particular
object or space is all to do with how the constituent components are
related to one another within the general framework of the assembly. In
the design and construction of a building, the civil engineer, the
architect, the mechanical engineer and the other professionals
coordinate to ensure that the objects and spaces are assembled in
accordance with the desired outcome. The process of assembling objects
and spaces is concerned with the fundamental concept of relationships
between entities. The study of such relationships is called topology.
Some examples of topological relationships are above, below, on top of,
under, adjacent to and inside.

Recently, researchers in robotics and CAD have emphasized the
importance of topological relationships between objects (Latombe, 1979;
Eastman, 1980). In robotics, some command languages have been

* Assistant Professor, Department of Art and Architecture, and
  Institute for the Study of the High-Rise Habitat, Lehigh University,
  Bethlehem, PA 18015.

developed that deal with topological relationships between components.
An example is the robot command language RAPT which was developed at
the Department of Artificial Intelligence of the University of Edinburgh
(Cameron, 1982). RAPT supports assembly line tasks whereby the required
relationships between shapes are anticipated by the programmed robot.

Topology-1 is a knowledge based CAD system that contains knowledge
about topological relationships between entities and knowledge about
geometry and various attributes of objects and spaces (Akiner, 1984).
It has reasoning capabilities about objects and spaces in buildings.

KNOWLEDGE ENGINEERING AND REASONING ABOUT OBJECTS AND SPACES

Knowledge engineering is a sub-field of artificial intelligence
that is concerned with the art of creating computer programs via
employing appropriate technique for knowledge representation, knowledge
deployment and knowledge acquisition. It is a field which provides
techniques that handle ill-defined problems in a knowledge-rich
environment. It also provides means to tackle problems which cannot be
solved by using algorithmic approach of programming.

In the knowledge engineering medium, symbolic models and symbolic-
inference making methods are implemented in order to represent and
manipulate knowledge. The entities that are dealt with in this
environment are "recognized", and the relationships between the entities
are "known" by the computer programs. In knowledge engineering, the
control structure of a system is totally separate from the knowledge
embedded in the system. In the process of reasoning about objects and
spaces, the knowledge about geometry and topology is represented
explicitly.

AIM OF TOPOLOGY-1

The aim of topology-1 is to develop a prototypical core element for
an interactive, knowledge based CAD system to support the design
activity of a designer. Topology-1 in its present development state is
capable of making inferences about topological relationships, geometric
entities, attributes of objects and spaces and provides support for the
designer in design appraisal. It is a system which incorporates
declarative attributes, explicit knowledge representation and
inferencing capabilities.

Topology-1 therefore embodies some of the characteristics of
knowledge based systems which most likely will be part of the CAD
systems of the future.

ASSUMPTIONS

In Topology-1, all components are parallelepipeds. This geometric
assumption implies that all angles are 90° and that all objects are
convex polyhedra. Thus, box geometry concepts are adopted as in
OXSYS/BDS (Hoskins, 1977). The descriptive homogeneity of orthogonality
is especially advantageous. In addition to the efficiency from the
inference-making point of view, this assumption will also require a
minimum number of facts to be stored.

CONFIGURATION OF TOPOLOGY-1

Topology-1 is composed of five components as shown in Figure 1 (Akiner, 1985). They are:

(i)   Knowledge (Rule) Bases
(ii)   Facts Bases
(iii)   Knowledge Base Library
(iv)   Facts Base Library
(v)   Attribute Library

These five components are linked to the "System Control" which is made up of the following elements:

(i)   Input-Output Operations
(ii)   Space and Object Creation Facilities
(iii)   Queries
(iv)   Reasoning by Inferencing
(v)   Pattern Matching (a form of inferencing)

In the following section a brief description of each of the above members of the Topology-1 configuration will be introduced.

Knowledge Bases

General Description of Knowledge Bases. A knowledge base is a set of computer files which contains knowledge (rules) about a particular field. In Topology-1 knowledge bases are mainly associated with topologies and geometries of objects and spaces and circulation within buildings.

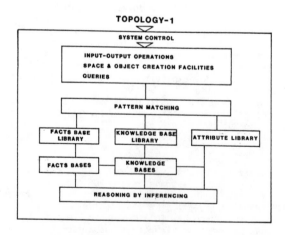

Figure 1   System Control and Components of Topology-1

Topology-1 contains 62 rules that are built into 8 different knowledge bases. In the first knowledge base the rules are concerned with low level knowledge such as set membership, finding minimum and maximum values, matching, finding absolute difference etc. The remaining 7 knowledge bases are registered in the knowledge base library and are called up as needed during the reasoning process.

<u>Types of Knowledge Contained in Knowledge Bases</u>. The knowledge bases of Topology-1 contains various types of knowledge (Gero, Akiner, Radford, 1983):

A. Knowledge About Individual Objects. Knowledge about the structure of a single object can be broken into three classes:

1. Knowledge about object topology. There are 9 topological relationships associated with general polyhedra (Gero et al, 1984). They are given as follows in the form X:Y (determine X given Y).

(i)	face	: face
(ii)	face	: vertices
(iii)	face	: edges
(iv)	vertex	: faces
(v)	vertex	: vertices
(vi)	vertex	: edges
(vii)	edge	: faces
(viii)	edge	: vertices
(ix)	edge	: edges

In traditional systems knowledge about topology is usually stored as algorithmic procedures. However, Topology-1 where logic programming has been used, provides both forward and backward induction facility which allows the representation of X:Y and Y:X in a uniform syntax.

2. knowledge about object geometry.
   length, width, area, volume.

3. knowledge about object attributes
   color, texture, material, price.

B. Knowledge About Relationships Between Objects.

1. topological relationships
   next to, between, on top of.

2. geometrical relationships
   bigger than, longer than.

C. Knowledge About Groups of Objects. In this class of knowledge, the focus of concern is on knowing if a group of objects has a particular meaning. For example the group of objects might be assembled to create a frame, a cantilever or a space.

D. Knowledge About Individual Spaces. A single space can be thought of as being similar to a single object except that it is hollow

and bounded by walls, floor and roof objects, so the same structural
characterizations apply as for single objects with the addition of a
wide range of attributes.

E. Knowledge About Relationships Between Spaces.

1. topological relationships
   adjacent to, under.

2. geometrical relationships
   smaller than, wider than.

3. circulation network relationships
   connected to, linked to.

F. Knowledge About Relationships Between Groups of Spaces. Buildings
are compositions of groups of spaces. Therefore knowledge about such
groups should be created and stored as part of knowledge about buildings.
As with groups of objects, the meaning of a group of spaces is sought.
This is knowledge which is domain independent, that is it does not
depend on the particular building in question, but applies to all
buildings which can be characterized this way.

Structuring and Contents of Knowledge Bases. In Topology-1
knowledge is built within a hierarchical framework as shown in Figure 2.
In this figure the interdependence of rules to one another is clear.
Beginning from basic values, definitions and relationships between
entities, higher level concepts are structured.

Each knowledge base contains a set of interrelated rules that are
defined in English and encoded in PROLOG (Clocksin and Mellish, 1981).
Consider the rules "edge", "higher than", and "above", in both
English and pseudo-predicate logic.

"An edge of an object A is a connection between two vertices if the
two vertices belong to object A and the two vertices are not
identical and the two vertices have exactly two coordinates of
equal value."

```
edge(V1, V2, Object) if
vertices(Object, [L]) and
member(V1, [L]) and
member(V2, [L]) and
V1 /= V2 and
coord(V1, [X1, Y1, Z1]) and
coord(V2, [X2, Y2, Z2]) and
match-two([X1, Y1, Z1], [X2, Y2, Z2]).
```

Note that here [] denotes a list and L represents a list of vertices.

"If A has a bottom face whose vertices have z coordinate values
that are $\geq$ the z coordinate values of the vertices of the top face
of B, then A is higher than B."

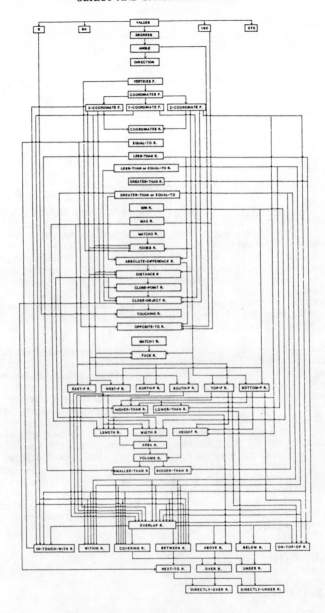

Figure 2   Relationships Between Topological Attributes
and Basic Definitions

```
higher than(A, B) if
top face(B, Z2) and
bottom face(A, Z1) and
Z1 >= Z2.
```

"If A is higher than B and A overlaps B along z axis, then A is above B."

```
above(A, B) if
higher than(A, B) and
overlaps(A, B, Z_axis).
```

Facts Bases

Topology-1 consists of two types of bases:

Initial Facts Base. It contains facts that are input to the system as a priori facts. Typical initial facts are associated with attributes of objects and spaces, particularly those that are based on the preferences of the designer and the client. These attributes may be material, shape, color, texture, price and so on. Some attributes are flexible such as price and some are fixed such as prescribed colors or textures. Hence, flexible attributes might need more frequent updating than fixed attributes.

Consider the following fact both in English and in pseudo-predicate logic: "The material of the frame of the building B is steel"
material(frame, B, steel).

Inferred Facts Base. This is a base that incorporates facts that are inferred by combining some knowledge with some facts. This base is built up as the system is run and inferred facts are accumulated via queries that are directed by the user. Each facts base contains a set of interrelated facts. Inferred facts are handled in a similar fashion as initial facts from storage and manipulation points of view.

In Topology-1, when the reference point, length, width and height of a particular object (or space) is input to the system, the system automatically infers the complete set of coordinates corresponding to the eight vertices of that object (or space) using the knowledge contained in the "space and object creation facility" of the system. Thus, an inferred facts base containing coordinates of vertices is created. This is the only inferred facts base within each domain which is not created by posing a specific query on the system. Some examples for the structure of such an inferred facts base in pseudo-predicate logic are as follows:

```
vertices(space name, [V1, V2, V3, V4, V5, V6, V7, V8]).
coord(V1, [X1, Y1, Z1]).
coord(V2, [X2, Y2, Z2]).
adjacent to(ColumnA, ColumnB).
on top of(BeamB, ColumnC).          etc.
```

Knowledge Base Library

In Topology-1, the system searches for knowledge base file numbers

in what is known as a knowledge base library. This could be viewed as
a process of searching for some specific item of knowledge as one would
search in a library. The knowledge base library contains all the
necessary meta-knowledge, that is knowledge about which knowledge is
encapsulated in which knowledge base. Meta knowledge is structured as
shown below:

    kb(_, knowledge base number, list of rules).
    e.g.: kb(_, know_base150, [above, below, covering]).

The first argument in the above structure represents the domain of
the knowledge. In this case the anonymous variable denoted by '_'
indicates that the particular knowledge that is encoded is domain
independent.

Facts Base Library

    Topology-1 consists of a facts base library which is similar to the
knowledge base library. The facts are always domain dependent by their
nature, in contrast to knowledge. Therefore the domain is always
specified in the declared facts about the placement of facts such as:

    fb(domain name, facts base number, list of facts).
    e.g.: fb(hotel, f_base250, [in_contact_with, adjacent_to,
                        on_top_of]).

Attribute Library

    The attribute library contains attributes of relationships. There
are three classes of relationships:

    Reflexive Relationships. If there exists a relationship between A
and B and the same relationship exists between B and A, then this
relationship is called a reflexive relationship. This rule which
encapsulates the piece of knowledge about reflexive relationships can be
expressed in pseudo-predicate logic as:

    reflexive(relationship) if
    relationship(A, B) and
    relationship(B, A).

Some examples of reflexive relationships are opposite to, touching,
next to, welded to.

    Antonymic Relationships. These are relationships that express some
form of an inverse position or opposite concept between entities. A
rule describing an antonymic relationship can be expressed in pseudo-
predicate logic as:

    antonymic(relationship1, relationship2) if
    relationship1(A, B) and
    relationship2(B, A).

Some examples of antonymic relationships are above-below, supports-
supported by, and north of - south of.

Transitive Relationships. Transitive relationships are
relationships that express some form of transitivity between attributes
or characteristics between entities. A rule describing a transitive
relationship can be expressed in pseudo-predicate logic as:

    transitive(relationship) if
    relationship(A, B) and
    relationship(B, C) and
    relationship (A, C).

Some examples of transitive relationships are above, below, etc.
Most of the transitive relationships are comparative in nature such as
smaller than, longer than, darker than and so on.
The declaration of the class of relationships is useful in a
knowledge based system because this facility reduces the size of
inferred facts bases. Furthermore, such a system proves to be more
efficient as these attributes of relationships when consulted decreases
the number of costly inferences to be made.

Input-Output Operations

These operations are associated with the encoding of a routine that
allows the input of sentences. In relation to this are some auxiliary
encodings such as reading in a single word given as initial character,
characters that form words on their own, words that terminate a sentence
and the output of inferences that are made.

Space and Object Creation Facilities

This facility allows the user to create spaces and objects. For
each space or object a name, X, Y, and Z coordinates of a reference
point and the length, width and height are provided by the user. For
each domain separate files are created that contain inferred facts
about vertices and coordinates of a certain space or object.

Queries

Queries in Topology-1 are concise questions about the following
aspects of objects and spaces:

Topologies. In Topology-1 a typical query is associated with
four parameters; Domain (D), Relationship (R), Object or Space A (H),
and Object or Space B (T). These parameters may be variables (var) or
nonvariables (nonvar). The number of queries resulting from the
combination possibility of these four parameters (D,R,H,T) is 2^4 . Out
of these sixteen possible queries only eight are found to be of interest.
Facts are stored in facts bases in the following generic form:

    $[D,R] \longrightarrow [H,T]$

When a query is posed on the system, it is read in by the use of a
predicate called "option". The query which contains key words is tested
by means of conditions of the predicate option and the flow is thus
controlled. An example of an option might be:

```
option(what, is, R, T, in, D, '?'):-....
                                  ....
```

Here the actual query might be 'what is on top of column C in office building?' which means a structure of the form:

```
[office building, on top of]—→  [H, columnC]
nonvar              nonvar       var nonvar
```

<u>Quantities</u>. The structure of the query is similar to the ones about topologies. Some of the typical queries are as follows:

How many spaces are there in domain D?
How many classrooms are there in domain D? etc.

<u>Circulation Networks (Paths)</u>. The query about circulation networks are also structured similarly to the previous queries. Some examples of queries are:

Through which spaces does one has to pass in order to go from space A to space B in domain D?
Through which spaces does one has to pass in order to reach from space A to space B taking the shortest route in domain D?

<u>Names of Objects and Spaces</u>. This query extracts the list of all objects and/or spaces in a certain domain. For example:

What are the names of spaces in domain D?

Reasoning by Inferencing

Any query that is posed by the designer invokes the appropriate mechanisms at the system control as shown in Figure 3. Firstly, the facts base library is loaded. The key predicate (word) which may be associated with topological attributes such as "on top of", "under", or may be associated with quantities such as "many" or perhaps may have to do with circulation that is expressed by the key predicate "reach". Here the key predicate expresses some relationship between two or more objects and/or spaces. The list of predicates that a particular facts base which is related to a particular domain contains is searched to see if the key predicate exists. If it does, then that particular facts base is loaded. If the inferred fact that satisfies the query posed exists in this facts base, then the appropriate answer to the query is given along with a message which tells where this fact is already stored.

In case the above described search within the facts base fails, then a new routine is invoked. The facts required to be combined with the knowledge base are identified and loaded. The key predicate is searched for within the predicate lists of different knowledge bases. If the searched-for knowledge is found, the conditions of rules are searched in facts bases where they might exist as inferred facts. All the inferred facts are used for satisfying the conditions. The conditions which are not satisfied in this process are re-evaluated using knowledge which is combined with the facts just loaded. New facts are inferred and are stored in appropriate facts bases. This process of inferring and storing the inferred facts is a recursive process and repeats until the requested query is answered.

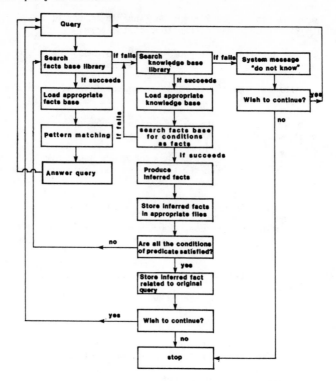

Figure 3   Schemata of Functioning of System Control

In case the above described process fails to satisfy a requested goal, then a third routine is invoked which is a message that the answer to the query is not known by the system.

Thus, in this approach, new facts are inferred as needed and a useful set of inferred facts is built up during a query session. These inferred facts can be fully or partially deleted by the user if need be.

Pattern Matching

This is a form of inferencing. In PROLOG which is a language where symbolic representation and manipulation is extensively used, this type of inferencing is a built-in feature of the language.

Reasoning About Spaces in Frank Lloyd Wright's Carlson House

Figure 4a shows tha plans for the three levels of Wright's Carlson House, built in Phoenix, Arizona in 1951 and Figure 4b shows the abstracted spaces with their associated names and floor levels. Queries

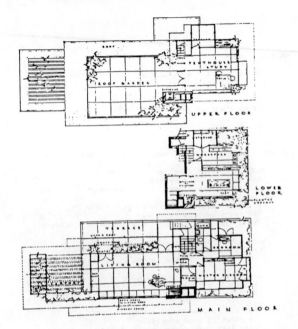

Figure 4a   Plans of Raymond Carlson House
                 by Frank Lloyd Wright

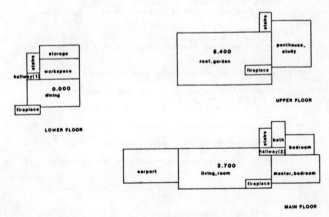

Figure 4b   Abstracted spaces with their names and
                 floor levels for Raymond Carlson House

about topology, quantities, circulation network and names of spaces about Carlson House are shown in Figure 5 and Figure 6.

wright : where is 440 length_l bath?

    440 is length_l bath in home.
    (this fact about home is already stored in f_base130.)

wright : is shape_of bath cube in home?

    no.

wright : is what is same_level dining in home?

    storage is same_level dining in home.
    (is found in know_base130.)
    (this fact about home is now stored in f_base140.)

    workspace is same_level dining in home.
    (is found in know_base130.)
    (this fact about home is now stored in f_base140.)

    stairs is same_level dining in home.
    (is found in know_base130.)
    (this fact about home is now stored in f_base140.)

    hallway(1) is same_level dining in home.
    (is found in know_base130.)
    (this fact about home is now stored in f_base140.)

    fireplace is same_level dining in home.
    (is found in know_base130.)
    (this fact about home is now stored in f_base140.)

wright : is carport longer_than bath in home?

    900 is length_l carport in home.
    (this fact about home is already stored in f_base130.)

    440 is length_l bath in home.
    (this fact about home is already stored in f_base130.)

    yes.
    (is found in know_base100.)
    (this fact about home is now stored in f_base150.)

wright : what is on_top_of living_room in home?

    roof_garden is on_top_of living_room in home.
    (is found in know_base150.)
    (this fact about home is now stored in f_base155.)

wright : what is above master_bedroom in home?

    penthouse_study is above master_bedroom in home.
    (is found in know_base150.)
    (this fact about home is now stored in f_base155.)

wright : is roof_garden covering living_room in home?

    yes.
    (is found in know_base110.)
    (this fact about home is now stored in f_base155.)

Figure 5   Queries

```
wright : what is over living_room in home?

    roof_garden is over living_room in home.
    (is found in know_base160.)
    (this fact about home is now stored in f_base160.)

wright : how many spaces are there in home?

    there are 14 spaces in home.
    (this fact about home is now stored in f_base185.)

wright : how many spaces are there in home?

    there are 14 spaces in home.
    (this fact about home is already stored in f_base185.)


wright : clear quantities about home.

    all quantitites about home are cleared.


wright : how do you reach from carport to penthouse_study in home?

    you reach from carport to penthouse_study by taking one of
    the following alternative routes:

    [[carport, living_room, master_bedroom, hallway(2),
    stairs, roof_garden, penthouse_study], [carport, living_room,
    master_bedroom, hallway(2), stairs, penthouse_study]]

    (this fact about home is now stored in f_base170.)


wright : how do you reach quickly from carport to penthouse_study in home?

    you reach quickly by taking the following route :

    [carport, living_room, master_bedroom, hallway(2), stairs,
    penthouse_study]

    (this fact about home is now stored in f_base175.)
```

Figure 6   Queries

IMPLICATIONS OF THE AVAILABILITY OF REASONING SYSTEMS

As shown in this paper, a great deal of inferences can be made by
providing the system with a limited amount of facts such as a
cartesian reference point, and dimensions of each object or space.
These facts coupled with the built-in knowledge allows the system to
infer novel facts and thus the system gets richer in terms of inferred
facts as the query session progresses.

The availability of reasoning systems such as Topology-1 alters the
future uses of computers in engineering and architecture. It changes
the computer from a dumb tool to a designer's aid with the capacity
to allow an individual designer or architect to encode his own
knowledge into it as well as gaining access to the knowledge of their
colleagues. Since much of this knowledge can be encoded declaratively,

(by stating 'what' rather than 'how') such systems produce a closer
contact with design goals without the present concentration of effort
on procedures. Reasoning systems will alter the emphasis in CAD away
from the simple drafting and elementary object modeling to an
elaboration of the immanent domain of buildings. This will make
computers a more useful aid to engineers and architects.

ACKNOWLEDGEMENTS

This work was supported by a grant from the Australian Research
Grants Scheme and a University of Sydney Special Project Grant. The
author is indebted to Dr. J.S. Gero and Dr. A.D. Radford of CARU,
Department of Architectural Science, University of Sydney for their
valuable critiques and encouragements. The author also acknowledges
Dr. L.S. Beedle of Lehigh University for his helpful suggestions and
comments.

REFERENCES

Akiner, V.T., "Topology-1: A System that Reasons About Objects and
Spaces in Buildings", Ph.D. Thesis, Computer Applications Research
Unit, Department of Architectural Science, University of Sydney,
1984.

Akiner, V.T., "Topology-1: A Reasoning System for Object and
Space Modeling via Knowledge Engineering, Proceedings of the 11th
Design Automation Conference, ASME, 85-DET-87, Cincinnati, Ohio,
1985.

Cameron, S., "A RAPT Picture-Scrapbook", DAI Working Paper, No.106,
Department of Artificial Intelligence, University of Edinburgh,
1982.

Clocksin, W.F. and Mellish, C.S., "Programming in Prolog",
Springer-Verlag, Berlin, 1981.

Eastman, C.M., "The Design of Assemblies", Institute of Building
Sciences Report, Carnegie-Mellon University, Pittsburgh, 1980.

Gero, J.S., Akiner, V.T., Radford, A.D., "What's What and What's
Where: Knowledge Engineering in the Representation of Buildings
by Computer", Parc83, Online Publications, London, pp.205-215,
1983.

Gero, J.S., Radford, A.D., Coyne, R., Akiner, V.T., "Knowledge-
Based Computer-Aided Architectural Design", IFIP, Knowledge
Engineering in Computer-Aided Design, Budapest, 1984.

Hoskins, E.M., "The OXSYS System" in Gero, J.S. (ed.) Computer
Applications in Architecture, Applied Science, London, pp.343-391,
1977.

Latombe, J.C., "Metodes et languages de programmation des robots
industriel", Seminaire International, IRIA, 1979.

KBES AND INTERACTIVE GRAPHICS

Vyomesh Vora,*M.ASCE

Abstract:

KBES is at research and testing stage at present.
Engineers have begun to accept Interactive graphics System
(IGS) as a tool in decision making. This paper discusses
information management in IGS and KBES environment for
decision making. KBES with IGS will enhance the acceptance
of KBES by Engineers. To achieve this objective, it is
necessary that engineers and researchers work together to
develop the requirements of KBES and understand the
constraints of KBES.

Introduction

Each of us to some degree consider ourselves experts
because of our experience and technical qualifications.
Engineers have been using computers extensively for analysis
and design of structure. In recent times, the use of
Interactive graphics System has assisted engineers to arrive
at a more rational understanding of impact of their
decisions. With the Knowledge Based Expert System (KBES) on
the horizon, we need to examine its implications and impact
on the role of engineers in fulfilling the responsibilities.

With the use of DBMS coupled with IGS, an engineer can
optimize the information management in the present IGS
environment. However, the practising engineers still feel
that IGS does not completely fulfill their needs for a total
information management. This is so, because many a times
such information is interdisciplinary and change in one set
of information impacts changes in the entire process of
design.

Will KBES in IGS environment help the engineer to
achieve fully integrated design process? Will he be able to
improve the heuristic approach of KBES through graphics
visualization?

Let us go through a design process in present
environment and determine what we will require to make KBES

*Product Specialist (civil/structural), CalComp 2411
W. La Palma Avenue, Anaheim CA 92801

more acceptable.

Responsibilities of an Engineer.

The responsibilities of an engineer, whose major effort is in structural engineering disciplines are mainly for Analysis and Design of structures. To achieve these objectives , an engineer is required to determine the layout of the structure, and subsequent to analysis and design phase, to develop enough 'detail' information so that the structure can be built per requirements.

Both, at pre and post processor stages, he has to take decisions on the various aspects of information. Preprocessing consists of managing information for defining the structure and at postprocessing stage, he checks the relevance of analysis based on his interpretation of analysis results to comply with the regulatory rules.

Information Management

The engineer mainly deals with the following sets of information.

--Specified Information

An engineer is primarily responsible for the decisions on specifying the materials of construction, size of the members, strength of materials used, to comply with the applicable code of practice and to ensure the overall safety of the structure against failures.

--'Derived' Information

More in Civil/Structural Engineering, than in any other discipline, the information about structure is derived from other disciplines.

For example, for an office building, the engineer ensures that the structural elements are compatible with architect's requirements for office space layout; or say, for a factory building, they are compatible with the machinery layout preferred by machanical engineers.

Another set of 'Derived' information is the analysis and design results.

--Repetitive Information.

Some of the information discussed above are of repetitive nature, because of the similarity of layout, loading or specific requirements to ensure uniformity.

Decision Making in IGS Environment.

We put more emphasis in visualization aspect, in using IGS as a decision making tool. We go through the iterative process of Analysis and Design until we reach to a conclusion in accordance with our comfort level and in compliance with regulatory rules.

Let us take an example of a beam element in a building.
To analyze and design, we go through the following steps.
--Locate beam in plan and elevation. (Assumption and visualization.)
--Specify material and assume size.
--Check clearances (visualization)
--Calculate loading and prepare model)
--Verify model (visualization)
--Analyze and Design check
--Verify size and clearances (visualization)
--Iterative process till user's requirements are met.

If this beam is part of a building frame, then when the frame is analyzed, the beam element is again checked iteratively for compliance with the user's requirements.

For optimizing the Building design the engineer studies the interface of information from various disciplines. Even though IGS environment helps engineer in arriving at a rational decision, he has to repeat the whole cycle of analysis and design for any change of information. To some extent, perhaps he can minimize the number of process cycles based on his experience and expertise.

Let us examine now, if engineers can fulfill their responsibilities more efficiently by use of KBES

Decision Making in KBES Environment

In KBES environment, the engineer will be able to optimize the design by using the opinion and knowledge of 'Expert' for decision making. The engineer puts more emphasis on rule based question answer aspect for decision making.

Let us take the example for beam element. Ideally, we would like the size of the beam to comply with all the constraints, such as clearance requirements, material capacity, and at the same time fulfilling all the requirements for a safe structure.

An expert can transform each of the constraint as a rule. He will devise the rules based on interdisciplinary information, with specific weightage to different rules according to his expertise.

The engineer can go through the forward chaining process in If-Then situation and arrive at a conclusion. In a situation where he knows the size of the beam element, he can go through the backward chaining process to ascertain whether all the constraints are fulfilled, resulting in an ideal design.

KBES is more of a heuristic approach to the problem assumptions and engineers have certain expectation of requirement for KBES. Let us examine these in brief.

Requirements of an ideal KBES

The task of building an expert system will include expertise from all the related disciplines for information management, so that any change in environment of one discipline is reflected by transfer of interdisciplinary information and revision rules.

The apprehension engineers may have for KBES is validity of expert's opinion. Engineers consider themselves as independent minded persons and may want to arrive at their own conclusion, or may want to check the validity of results based on expert's opinion. One way to overcome this is to develop KBES in IGS environment, so that a user can visualize on screen the implication of expert's decision at various stages of design, and make changes in interactive manner to assert his expertise and opinion.

Will the KBES be ideal in all the circumstances? How much 'expert knowledge' we can compress in one system? How do we compare one expert system against another? How do we ensure the cost effectiveness of the system, if the rules are based on regulatory criteria and changes in the regulations may affect the validity of the decisions?

To ensure the acceptable system, we need to think

about these questions.

Conclusion.

We have seen that the engineers have reached at a comfortable level with the analysis and design in IGS environment, because it gives them full control of decision making based on their input.

To reach a level of comfort for use of KBES, it is necessary that engineers have input and idea about KBES. The Appendix lists books which will give details about KBES and its development process. Acceptance of KBES will be based on how much input engineers have for development, and an individual's knowledge about the nature of Expert Systems.

This is an ideal time to start dialog between researchers and engineers on the requirements and acceptance criteria, so that the resulting Expert Systems can be both useful and effective.

Appendix-References.

Hayes-Roth, F., Waterman, D.A., Lenat, D.B., (editors), BUILDING EXPERT SYSTEMS, Addison-Wesley, Reading, MA,1983

Schank, R.C., Childers, P.G., THE COGNITIVE COMPUTER ON LANGUAGE, LEARNING, AND ARTIFICIAL INTELLIGENCE, Addison-Wesley, Reading,MA,1984

Winston,P.H., ARTIFICIAL INTELLIGENCE, Second Edition, Addison-Wesley, Reading, MA,1984

Winston, P.H. Berthold K.P.Horn, LISP, Addison-Wesley, Reading, MA, 1983

APPLICATIONS OF AUTOMATED INTERPRETATION TO SENSOR DATA

Kenneth R. Maser*, Affiliate Member, ASCE

Abstract

Sensor data is being increasingly utilized by civil engineers to evaluate the condition and properties of materials and structures in-situ. The primary limitation to the full exploitation of available sensor technology lies in the ability to accurately and efficiently interpret the large quantity of data which can be generated. A knowledge-based concept for automating the interpretation of sensor data (AISD), presented elsewhere, offers the potential for overcoming this limitation. This paper focuses on potential applications of the AISD concept in civil engineering. A number of potential applications are proposed in the areas of site investigation, construction monitoring and control, and condition assessment. The potential value of the AISD approach is assessed, based on a number of generic features, for each potential application. One application, the high speed sensing of bridge deck deterioration, is developed as a prototype example. Discussion of the knowledge-base and computational tools developed for this example illustrates a basic approach which can be used to address other civil engineering applications.

Introduction

The acquisition of accurate information concerning the physical condition of materials and structures in-situ is key to all aspects of Civil Engineering. This information supports the design, construction, operation, and maintenance of constructed facilities. In recent years there has been a greater emphasis on obtaining more, and higher quality data. This emphasis has evolved for a number of reasons. The increasing cost of construction (adjusted for inflation) has required that sites be better characterized before construction begins. The economic impact of unexpected conditions in applications such as tunnelling can be substantial. The emphasis on environmental protection has created a need to better understand and measure conditions under the ground. The need to identify, locate and track contaminant plumes is an example of such a need.

The growing demands of managing and maintaining our aging infrastructural systems have added a new dimension to the need for in-situ condition and deterioration data. The availability of accurate and comprehensive condition and deterioration data is becoming increasingly important to a number of aspects of facilities management, including:

*President, Civil and Environmental Technology Associates, Arlington, Massachusetts, and Research Associate, Massachusetts Institute of Technology, Cambridge, Massachusetts planning future maintenance expenditures and associated financing; deterrence of deterioration using low cost preemptive measures; developing a rationale for prioritizing multiple repair and rehabilitation projects; and minimizing cost overruns by accurately specifying the required type and extent of maintenance. The demand for this data is common to all infrastructure systems (e.g., water supply, transportation, and waste disposal and treatment) and their associated elements (e.g., buried pipelines, tunnels, dams, pavements, bridges, and track).

A number of sensory techniques have been developed over the years which have the potential for meeting these growing demands. These include laser optics, ground penetrating radar, infrared thermography, and electromagnetic conductivity. Such techniques bring powerful measurement potential to construction and civil engineering applications. This potential has yet to be fully realized for a number of reasons. One reason is that these techniques are sensitive to normal variations in environmental conditions (temperature, moisture, etc.) as well as to the specific condition that are to be measured (e.g., strength, density, etc.). Thus, it has always been a problem separating out the desired measurement from these uncontrolled environmental variations. A second reason is that far more data is generated by these techniques than can be interpreted manually. This data overload is aggravated in situations where one measurement technique is not adequate to fully assess the conditions of interest, and data from two or more measurement techniques is desirable. The sensory data itself usually contains a large quantity of potentially useful information which is often processed away to facilitate manual interpretation. Thus, information which may help to distinguish environmental variations from desired measurements is lost because there is too much of it to handle. A final reason is that in-situ measurements, such as condition and deterioration surveys, inherently require a large quantity of data, since the structures under analysis are extensive.

In order to extend the potential value of existing sensory techniques, a method to automate the interpretation of large quantities of sensory data (AISD) has been proposed (Maser, 1986). With this method, data simplification required for manual interpretation can be eliminated, thus preserving the full content of the sensory data. Furthermore, sensory data taken over large areas or distances can be interpreted automatically without time-consuming, costly and tedious analysis. The AISD concept achieves this automation by encoding the experience and judgment which would normally be required to interpret the sensory data. This knowledge is multidisciplinary, and covers the sensory instrumentation and analysis of sensor signals; the physical and material properties of the materials and phenomena that are being sensed; and the design, construction, and performance of the facility, structure, or element being evaluated. For example, interpretation of a sonic survey of a dam to evaluate concrete strength in-situ requires knowledge of the sources, receivers, and equipment utilized to collect the data; of the nature of the sonic waves that are generated and how these waves respond to conditions of interest in the concrete; and of the overall design and construction of the dam.

The AISD concept organizes the knowledge by discipline, encodes it, and automatically applies it to the sensor data utilizing knowledge representation and inference techniques developed in the field of Artificial Intelligence(AI). The content and structure of this knowledge, and the computational tools for automatically applying this knowledge to the sensory data are the central issues addressed in developing a workable system.

Background

Efforts to apply encoded knowledge to the interpretation of sensory data, as proposed herein, have been pursued in a number of applications. DENDRAL [Lindsay et al, 1981] was developed to predict molecular structure from mass spectrograms. SIAP [Nii et al, 1978, 1982] attempted to synthesize information from a variety of acoustic surveillance sources in order to identify enemy vessels. The DIPMETER ADVISOR [Smith, 1984] was developed to assist in determining geological conditions from dipmeter logs. One of the most extensive efforts in the field of knowledge-based signal processing is embodied in HEARSAY-II [Erman et al, 1980]. This project attempted to automate the interpretation of the human speech. The input to the system was acoustic signals representing human speech, and the output was to be a high level interpretation of the meaning of the speech. The project sought to integrate the numerous disciplines involved in this process (e.g., phonetics, syntactics, semantics, etc.) through the establishment of independent knowledge bases interconnected through a "blackboard". In comparison with previous work, the AISD concept looks like a simplified HEARSAY-II; i.e., a data-driven system to automate the interpretation of signals acquired in a highly variable measurement environment, using multiple knowledge sources. One common aspect of the various applications of knowledge-based signal processing which have been pursued to date are the features which defined the applicability of the knowledge-based approach. These can be summarized as follows:

1. Sensors produce signals and data which do not have an obvious interpretation.

2. Interpretive skills are required (usually from more than one discipline) in order to interpret the data in the context of the problem being solved.

3. The amount of data and/or interpretive steps that need to be applied exceeds the capacity of the individuals capable of doing this interpretive work.

4. An accepted body of interpretive knowledge exists to serve as a source which can be encoded and automatically applied.

5. The resulting system, if successfully developed, would achieve performance that could not otherwise have been achieved, or could match existing capabilities at considerably lower cost.

Applications that do not combine these features will be more efficiently handled using conventional algorithmic techniques. For example, if signals are not obvious, but can be clarified and easily interpreted after a number of computations have taken place, then digital signal processing should be adequate and efficient. Analysis of previous research (Davis, et.al, 1982) has shown that it is necessary to use a problem-oriented approach rather than a discipline oriented approach (e.g., AI vs. DSP) when addressing a given measurement and interpretation application. In this context, the thrust of this paper is to define a number of sensory measurement applications of Civil Engineering interest, evaluate the applicability of the AISD concept to each application, and develop a specific application to demonstrate AISD implementation.

Description of the AISD Concept

The AISD concept is described in detail elsewhere (Maser, 1986), and is summarized here for clarity. The AISD system takes electronic signals generated by one or more sensors, combines information from these signals with information provided by the user, and produces interpretation of in-situ conditions in language understandable to the user. Analog and digital processing of the sensory signal are kept separate from the knowledge-based processing. The knowledge-based system simultaneously integrates sensory data, user input, and encoded knowledge. The system has a modular capability to accept input from two or more sensors. It is data driven, and has the capability of explaining how a conclusion was reached. Data comes from two sources. Electronic data (usually as a voltage) arrives from the sensors. User data characterizing the measurement application is input manually by the user or automatically from the user's database. The electronic signals are initially processed using conventional analog and digital processing techniques (A/DSP), and signatures are determined. The selection of processing techniques and signature algorithms is be guided by the knowledge-based program. The computed signatures are passed to the knowledge-based program for interpretation. This interpretation is carried out in conjunction with other information provided by the user. Knowledge is organized in four generic disciplines with the conclusions of one knowledge base providing input to another. The knowledge is represented as a set of if-then rules.

Sensor signatures provide input to the knowledge base which converts signatures to symbolic descriptions (KB4). Specifically, this knowledge base takes numbers as input and produces verbal descriptions (e.g., "high attenuation", "possible scattering", "low velocity", etc.) as output. These symbolic descriptions are passed on to KB3, which interprets these descriptions in terms of physical conditions of the material or structure being sensed. Note that each sensor, with its associated A/DSP, signature algorithms, and KB3/KB4 combination, constitutes a separate module. The presence of each such module enhances the performance of the system by providing more information. The absence of any particular module does not, however, prevent the system from operating. On the other hand, KB1 and KB2 are generic to the application, and have little relationship to the sensors. The input to KB2 comes directly from the user data. Some of this data, such as design and inspection data regarding a particular structure, may be retrieved automatically if this information is accessible through a database. KB1 is the knowledge-base which assembles all of the supporting information into a framework which describes the phenomenon or property to be measured. In any application KB1 may be a complex combination of analytic solutions, empirical regressions, and general rules

of thumb. KB1 absorbs whatever information is made available to it by
the user and the other knowledge bases, and produces conclusions which
are finally passed to the user.

Review and Evaluation of Potential Applications

     In order to evaluate the applicability of the AISD concept, a num-
ber of in-situ measurement concepts and applications which might be
suitable to knowledge-based processing are proposed. These concepts and
applications are evaluated according to their suitability, and one
amongst them is selected to serve as a demonstrative example. The
proposed applications are not intended to be comprehensive or exhaus-
tive, but rather to highlight potential applications, and to develop a
rationale for applying the AISD approach.

     The scope of in-situ measurement applications considered here in-
clude those conducted strictly for informational purposes, as well as
those which provided a basis for real time feedback and control of construction and managment operations. All phases of the life cycle of
constructed facilities have been considered, including design, construction, operation and maintenance, and repair and rehabilitation. In-situ
data requirements supporting design are usually those associated with
site characterization. Those supporting construction include construction monitoring, quality control, and feedback and control of equipment
and processes. In-situ data is required during operation and maintenance in the form of condition surveys and performance monitoring.
The same type of data is required for repair and rehabilitation, but
with a greater level of detail in order to support specifications for
repair and rehabilitation projects.

     A number of general concepts and applications under the headings
listed above are proposed and described in the following paragraphs.

     Site Characterization

     a) Tunnelling
The a priori prediction of underground conditions (rock strength
and quality, faulting, stratification, aquifers, lenses, etc.) constitutes a critical aspect in the design and construction of a tunnel.
Millions of dollars, and lives, are at stake when unexpected underground
conditions are encountered during a tunnelling project. Therefore, considerable effort has been invested in the development of sensory tools
to predict underground conditions more thoroughly and accurately than is
typically possible through a set of borings [FHWA, 1979]. These efforts
have focused on electromagnetic, seismic, and acoustic emissions
techniques. The use of these sensory techniques have not been fully accepted by the tunnelling design and construction community due to their
cost and uncertainty. The proposed AISD concept would involve integrating the information generated by these sensory methods with that
generated by traditional geotechnical techniques. This integration
would enhance the certainty of the prediction.

     By combining knowledge and data sources, the proposed AISD applica-
tion could also reduce the total data requirement, and perhaps lower the
cost of the survey. Ultimately, the knowledge base could incorporate

the various choices of design and and construction techniques, so that the detail generated would match the detail required by the user to make decisions.

b) Hazardous Waste Site Characterization

Attention to this area is relatively recent, and numerous measurement tools have been developed and are being adapted for this purpose. Such tools include geophysical techniques such as subsurface radar, electromagnetic conductivity, electrical resistivity, and seismic techniques for geological characterization, plume detection, and barrier evaluation; chemical analysis techniques such as portable gas chromatographs and vapor analyzers, and downhole chemical analyzers; groundwater flow monitoring techniques, such as downhole flowmeters; and conventional geotechnical data. The combination of extensive data from varied sources, coupled with the variety of disciplines traversed, has made hazardous waste site characterization a costly, and often inconclusive process. The proposed AISD technique would integrate these diverse sources of information in order to fully assess the potential hazard posed by particular site, as well as to serve as a basis for selecting the appropriate remedial action. Such an integration would combine expertise in geophysics, hydrology, analytical chemistry, geotechnical engineering, and environmental engineering.

c) Dams

Characterization of abutment and underlying material associated with a dam is critical in designing a structure which will have adequate structural and hydrological stability. The consequence of inadequate design or of unexpected conditions can be catastrophic. As in tunnelling, combinations of geophysical and geotechnical techniques are utilized. The proposed technique, therefore, would be similar to that proposed in the tunnelling application discussed earlier.

Construction Monitoring and Control

a) Monitoring of In-Place Density and Moisture

A number of construction processes involve the placement of material to a specified density and moisture content. These include soil compaction, placement of compacted clay liner, placement of roller compacted concrete, and the continuous slipforming of pavements. Real-time feedback indicating that density and/or moisture content is out of spec can be used to correct construction conditions. This correction can prevent the much more costly rebuilding which may be required later on.

In-situ density is usually measured statically using a nuclear backscatter technique. Moisture content is usually measured by taking samples to the lab. The proposal here is for a mobile, field technique for simultaneously measuring density and moisture content combining nuclear and an electromagnetic technique. The interpretation of this multisensor system would be subjected to environmental and application related conditions, and would be assisted by a knowledge-based processor.

b) Obstacle Avoidance during Excavation

In many construction, repair, and rehabilitation applications it is important to locate and work around natural or man-made obstacles such as boulders and underground utilities. Inability to accurately locate these obstacles increases construction time dramatically, and results in damaged utility lines, damaged construction equipment and worker safety hazards. The concept here would be to equip construction machinery with sensors which can locate and characterize these buried objects, and which can organize this data in a manner which provides input into the guidance of the equipment. This concept has been pursued for buried metal objects such as pipelines [Whittaker et al., 1985] using conventional algorithmic feedback and control techniques. As the characteristics of the object become less dramatic (e.g., plastic pipe vs. metal), the need to assemble more sources of knowledge will be greater, thus suggesting a knowledge-based approach.

c) Interactive Construction

Although not commonly employed in the U.S., there are certain construction processes which are carried out and controlled according to measurements made during the construction process. One such technique which has been recently introduced in the U.S. is the so-called "New Austrian Tunnelling Method" (NATM). A number of measurements are made of rock movement and deformation during construction, and ground support and excavation sequence can be carried out in response to these measurements. For such applications it would be desirable to integrate more sources of data, and more knowledge, in a manner which could accelerate the construction process. This function could be served by the proposed AISD knowledge-based system.

d) Excavation of Buried Objects

Obstacle avoidance was mentioneed earlier. Here, consideration is given to applications where a buried object is to be removed. Such applications include situations where direct human intervention might be dangerous, such as excavating gas lines or removal of drums at a hazardous waste site. The sensory methods used must both locate the object and characterize it with sufficient accuracy so that mechanical devices can support it. This is more complex than the obstacle avoidance problem. On the other hand, object location, size, and shape are often known to some degree. Therefore, this prior information might be usefully employed through a knowledge-based approach.

Condition Surveys

a) Pavements

In the U.S. there are four million miles of pavement being maintained at an annual cost of over a billion dollars. One rational basis for budgeting and allocating these dollars is an assessment of the condition and remaining life of the pavement. Measurement technology for obtaining pavement condition data at highway speed is essential in order to provide accurate, current, and comprehensive data. Progress in this area includes high speed profilometer and ridemeter devices, and optical techniques have been under development for the high speed evaluation of texture and surface distress. Technological development in this area has been limited by a traditional one-sensor/one-measurement approach. This approach has been somewhat dictated by the kind of

measurement data that state highway agencies have traditionally utilized
in their management systems.

The proposal here is for a measurement system that integrates the
data from a "suite" of sensors with empirical, theoretical, and judgmental knowledge related to pavement behavior. For example, the suite of
sensors might include input from optical imaging at the pavement surface, ground penetrating radar, and surface profilometers. These three
sensory inputs provide data points to the pavement condition picture, a
picture which could be tied together with encoded knowledge drawn from
30 years of research into pavement behavior.

b) Bridge Decks

Thousands of bridge decks are prematurely deteriorating due to
corrosion-induced spalling. The conditions that lead to spalling
(corrosion, cracking, and delamination) are subsurface and are not apparent to the maintenance organization until spalling actually occurs.
A number of techniques have been developed to evaluate these conditions,
but most are slow, labor intensive, and require lane closures. Recently,
high speed sensory techniques such as ground penetrating radar and infrared thermograhy have been explored [Joyce, 1984]. These techniques
have yet to fully establish reliable and credible assessments of bridge
deck consditons. One reason is that the data from these techniques have
been looked at independently, and have not been analyzed in the context
of other bridge deck knowledge.

The proposal here is for a technique which analyzes combined radar
and infrared signals in the context of knowledge of the bridge deck environment and of the bridge deck deterioration process. Considerable
research has been conducted over the past 15 years to support the latter
knowledge base.

c) Subsurface Leakage Detection

Subsurface leakage occurs in water pipes, storage tanks, and waste
impoundments. At minimum, it results in the loss of a valuable com-
modity (e.g., water-utility losses up to 25% have been reported). At
worse, such leakage can lead to environmental hazards, as in the case of
leakage of toxic substances into the groundwater. The identification
and location of subsurface leaks is a difficult measurement problem
either because the quantity of the leak is small compared to the total
volume of liquid (as in an impoundment), and/or the leakage is distributed over a very broad area. Leakage must therefore be inferred
indirectly through subtle changes in the facility or the surrounding environment [Maser, et al.]. The application of a knowledge base could
enhance this inference, especially if the facility is extensive, or if
there is a large number of them. Encoded knowledge could incorporate
information about surrounding soil properties, leakage history and
design of the particular containment structure, etc., and apply this information in an organized manner to the sensor data.

d) Corrosion Detection

Corrosion often occurs in locations that are not easily accessible
to visual inspection. Such corrosion has lead to failures such as that
of the Mianus Bridge on I-95 in Greenwich, Connecticut, and that of a
number of building facades in New York City. It would be desirable to

develop techniques which can "see" corrosion when it is not visible. Such techniques may have to infer corrosion from other measurements, such as moisture environment, geometric distortions, etc. Once again, where inference is involved, support from a knowledge base would be useful. The specific sensory embodiment of such a technique would require further investigation.

e) Roofing Evaluation

Roofing installation and repair is a multibillion dollar industry in the United States. Development of techniques to rapidly locate and evaluate roof leaks would have a significant economic potential. The problem with a leaky roof is that the observation of a leak within a building often has nothing to do with the location of the source of the leak in the roof. Various sensory techniques such as infrared thermography and electrical resistivity have helped in locating the defective area in the roof. This data might be effectively combined with knowledge of the building type, construction method, construction materials, and layout of the HVAC and other building services. A knowledge-based approach could provide an effective tool for combining these information sources.

f) Dam Safety

The catastrophic consequences of a dam failure have made dam safety assessment an important issue. The diagnosis of the safety of a dam is usually based on a combination of inputs, including design data, condition data, and hydraulic load data. Traditionally, condition data has been limited to conventional geotechnical techniques. It may be possible to incorporate more comprehensive condition data from other sensory methods if a mechanism for integrating this information with other sources existed. Such an integration could be accomplished through a knowledge-based technique.

Evaluation of Candidate Applications

The various potential applications presented in the previous paragraphs were evaluated according to their potential value of a knowledge-based system for automated data interpretation. The evaluation factors were based on the criteria stated earlier. The evaluation itself was based on personal interviews, published reports, conference papers, trade literature, and the personal experience of the author. Table 1 summarizes the results of the evaluation. The evaluation factors shown in the table are restatements of those factors that were listed earlier. Ratings in the table that are "high" support the suitability of the knowledge-based approach, while those that are "low" weigh against it. The following discussion highlights some of the conclusions presented in the table, along with the rationale supporting these conclusions.

Under "obscurity of data" (factor 1), those sensory methods which could be strongly correlated with the desired measurement were considered "low", while those from which the desired information had to be inferred were considered "high". For example, the geophysical data generated for tunnel site investigation was considered "medium" because, while they require inference, they are localized to the tunnel alignment, and the inference is specific (water, strength). The geophysical

techniques used for hazardous waste site investigations were rated high
on data obscurity because the subsurface of interest has undefined
dimensions, and because more specific details (plume dimensions,
permeability) need to be inferred. Density monitoring was rated low in
this category since the correlation between nuclear measurement and density can be algorithmically computed and corrected. Under factor 2, the
level of interpretive skills required for density measurements is also
low. Most of the other applications rated high for this second factor
because a number of knowledge sources are required to integrate the data
into the overall measurement objective.

In terms of quantity of data and interpretive steps (factor 3) condition surveys all rated high since these surveys involve massive
inventories and extensive facilities. Hazardous waste sites rated high
because of the complex interdisciplinary interplay amongst a wide
variety of data sources. The most distinguishing factor is (4), the
quality of knowledge sources. This factor makes a clear distinction between a relatively new field with a limited body of accepted and
documented knowledge (e.g., hazardous waste site investigation, leak
detection), and a field that has been the subject of many years of organized and documented study (e.g., pavements, dams).

The fifth factor relates the knowledge-based approach to
alternatives. The ratings here are judgments which would require some
further study to substantiate. To illustrate the basis of the judgment,
the tunnel application was rated "questionable" because of the debate
within the industry regarding the value of higher quality data. Some
say that the effort should be directed towards "universal" tunnelling
machines and techniques which would be relatively insensitive to varying
ground conditions. All of the construction monitoring and control applications rated moderate because it was felt that good quality algorithms might be sufficient and provide the speed appropriate for real-
time feedback. The condition surveys all rated high because it was felt
that all required some type of judgment or experience, and all required
integrating knowledge from various sources.

Based on the evaluation summarized in Table 1, the bridge deck application has been selected for further consideration as a model to
exemplify a prototype AISD system as described below. In addressing the
bridge deck problem the sensory approach has been narrowed to the use of
ground penetrating radar. Radar produces a full waveform embodying
several pieces of information. Thus, considering the full radar
waveform is equivalent to taking input from more than one sensor. The
subsequent integration of infrared thermography is straightforward.

Demonstrative Example - Bridge Deck Condition Surveys

In order to demonstrate the application of the AISD concept, the
problem of high speed bridge deck evaluation has been selected as a representative example. R&D efforts in this area have shown consistent
positive indications that ground penetrating radar can detect
delaminated conditions. No rational physical model has been proposed to
explain the various observed responses, and different investigators have
identified different interpretive signatures. Data interpretation has
been primarily manual and subjective. Results have been correlated to

EXPERT SYSTEMS IN CIVIL ENGINEERING

TABLE 1 - EVALUATION OF POTENTIAL APPLICATIONS

Evaluation Factor / Potential	1. Obscurity of Raw Data	2. Level of Skill Required	3. Quantity of Data/Inter-Pretive Steps	4. Quality of Inter-Preting Knowledge Sources	5. Potential Performance/ Cost-Benefit vs Alternative Approaches
1. Site Investigation					
Tunnels	moderate	high	moderate	high	questionable
Hazardous Waste Sites	high	high	high	low	high
Dams	moderate	high	moderate	high	moderate
2. Construction Monitoring & Control					
Density Monitoring	low	low	high	high	moderate
Obstacle Avoidance	moderate	moderate	moderate	low	moderate
Interactive Construction	moderate	high	moderate	high	moderate
Excavation of Buried Objects	moderate	high	high	moderate	moderate
3. Condition Surveys					
Pavements	moderate	high	high	high	high
Bridge Decks	high	high	high	high	high
Subsurface Leakage Detection	high	high	high*	low	high
Corrosion Detection	high	high	high	moderate	high
Roofing Evaluation	high	high	high	low	high
Dam Safety	moderate	high	high	high	moderate

*in those applications of interest

existing delaminations, but not to any other subsurface conditions which
precede delamination. Prior knowledge of the bridge deck deterioration
process has not been employed in interpreting the data. Overall, the
results have not been clear and convincing, and there is no widespread
acceptance of the technique. Therefore, this application was selected
as an example with the belief that the AISD concept applied to the radar
technique could enhance its basic potential. The development of this
application consisted of the following steps: development of the
knowledge base; development of computational tools; and conduct computational experiments.

The knowledge base is broken down into four areas by source, as
follows:

KB1. <u>Bridge Deck Deterioration</u> This knowledge base represents all of
the processes and relationships that characterize the development
of spalling in bridge decks. These range from heuristics such as
"freeze-thaw cycles promote the growth at cracks," to empirical
relationships such as those which have been established between
number of salt applications, rebar cover, and corrosion initiation.

KB2. <u>Bridge Design, Construction, and Maintenance</u> This knowledge base
represents those aspects of bridge design and construction which
may have some bearing on the prediction and understanding of
spalling. It also includes knowledge of current inspection practices, and the relationship between the information generated by
these practices and the current condition of the bridge. This
knowledge base contains primarily heuristics, such as knowledge of
different design types, influence of design type on the deterioration process, and possible deviations from as-built drawings due to
construction techniqes.

KB3. <u>Radar Response to Bridge Deck Conditions</u> This knowledge base represents the physical response of pulsed electromagnetic energy to
the variety of conditions which may exist in a bridge deck. This
knowledge is represented as a set of heuristics which relate
various characteristics of the radar waveform to possible physical
conditions within the deck. Examples of such relationships are
connections between radar attenuation and concrete moisture content, and the connection between intermediate reflections and
cracking.

KB4. <u>Radar Signature Analyses</u> This knowledge base interprets the signature in terms of symbolic descriptions such as amount and
location of attenuation, appearance of reflections and scattering,
and actual vs. expected arrivals. The signatures represent the
simplified "fingerprint" of the waveform extracted through analog
and digital processing, and include such features as peak arrival
times and amplitudes, zero crossing times, and integration of
selected portions of the waveform.

The computational tools that were used in this application
included: a knowledge-based program using IMST (KNOWL); a signature
evaluation program (SIGNAT); a waveform synthesis program (SYNTH); and a
prototype intelligent bridge deck analyzer (IBDA). KNOWL represented

the software embodiment of the knowledge-based processor shown in Figures 1 and 2 and SIGNAT represents the A/DSP component. KNOWL uses an expert system "shell" (IMST) written in LISP, developed for generalized database management. SYNTH synthesizes raw radar data which can be analyzed in the absence of true field data. The IBDA integrates programs 1, 2, and 3 with a user interface in order to carry out a complete analysis of a bridge deck.

A total of twenty rules were constructed to represent the knowledge base. The user input incorporated into these rules includes the following elements for the bridge deck: years since previous reconstruction; a rating of the moisture environment; specified water/cement ratio of placed concrete; number of salt applications per year (estimated); a rating of the freeze-thaw environment; a rating of the traffic loading intensity; and the top-rebar cover from the as-built drawings.

Once the user data is entered, the program begins to read the radar waveform and SIGNAT computes signatures for each location. These are passed to KNOWL, which concludes by summarizing the key conclusions for all locations in a table as shown in Figure 1. The table lists locations (x,y), attributes, and values for these attributes. The attributes are likelihoods of delamination (dlam) and cracking (crck).

x	y	dlam	crck	corr	chlo	lcov	cover
1.0	0.0	0.00	0.08	0.12	0.11	0.00	2.00
2.0	0.0	0.17	0.34	0.41	0.18	0.90	1.50
3.0	0.0	0.00	0.08	0.12	0.11	0.00	2.00
4.0	0.0	0.00	0.08	0.12	0.11	0.00	2.00
5.0	0.0	0.12	0.25	0.12	0.11	0.00	2.00
6.0	0.0	0.17	0.34	0.41	0.18	0.90	1.50
7.0	0.0	0.17	0.34	0.41	0.18	0.90	1.50
8.0	0.0	0.23	0.46	0.41	0.18	0.90	1.50
9.0	0.0	0.00	0.15	0.12	0.11	0.00	2.00

Figure 1: Table of Conclusions

The prototype tools described above were utilized for computational experiments whose objectives were (1) to evaluate the application of a knowledge base to the automated interpretation of sensory data characterizing in-situ conditions; and (2) to demonstrate the integration of a complete system combining user interface, digital, and knowledge based processing. The significant findings from these experiment are summarized below.

1) Empirical relationships can be effectively combined with heuristics in the knowledge base, and used for data interpretation.

2) The weighting and certainty associated with different portions of knowledge have a strong impact on the use of information.

3) Similar conclusions derived from different sources reinforce one another.

4) The interpretation of sensory data is heavily influenced by what the user already knows.

Finding (2) above revealed the difficulty of dealing with uncertainty in the presence of combined numerical and heuristic knowledge, and numerical and qualitative input data. A hybrid system of certainty factor computation was implemented, but will require further consideration in the future. Findings (1), (3), and (4) exemplify the capabilities that were intentionally sought.

Conclusions

This paper has described how the application of sensor data in civil engineering can be enhanced by automating its interpretation. A number of candidate applications have been discussed, and the potential value of the AISD concept for automated interpretation has been described for each application. The characteristics of a given application which define the appropriateness of the knowledge-based approach have been presented, and the candidates have been evaluated based on these characteristics. This evaluation has shown that condition surveys represent a fertile area for the application of the AISD concept. The automated inspection of bridge decks for subsurface deterioration has been developed as a representative example. A discussion of this example has illustrated the steps required to implement the AISD concept, the typical content of a knowledge-base, and the software tools required for implementing a representative AISD application.

Acknowledgement

The author would like to acknowledge the National Science Foundation for their support of this work under an SBIR Phase I grant.

References

Alongi, A.V., T.R. Cantor, C.P. Kneeter, and A. Alongi, Jr., "Concrete Evaluation by Radar Theoretical Analysis," Transportation Research Record No. 853, 1982.

Chung, T., C.R. Carter, D.G. Manning, and F.B. Holt, "Signature Analysis of Radar Waveforms Taken on Asphalt Covered Bridge Decks," Canada Ministry of Transport and Communications, Report ME-84-01, June 1984.

Clemena, G., "Non-Destructive Inspection of Overlaid Bridge Decks with Ground Penetrating Radar," Transportation Research Record No. 899, 1984.

Davis, R., et al., "Knowledge Based Signal Processing," Trends and Perspectives in Signal Processing, July, 1982.

Erman, L.D., F. Hayes Roth, V.R. Lesser, and R. Reddy, "The Hearsay-II Speech Understanding System: Integration of Knowledge to Resolve Uncertainty." ACM Computing Surveys, Vol. 12, No. 1, pp. 213-253, June 1980.

FHWA, "Tunneling Technology for Future Highways." Department of Transportation Annual Project Report, Project 5B, September, 1979.

Joyce, Richard P. "Rapid Non-Destructive Delamination Detection," FHWA Report, FHWA/RD-84/1076, November, 1984.

Lindsay, R., et al, Applications of AI for Organic Chemistry: The DENDRAL Project. McGraw-Hill, New York, New York, 1981.

Manning, D., and F.B. Holt, "Detecting Deterioration in Asphalt Covered Bridge Decks," Transportation Research Record No. 899, 1984.

Maser, K.R. and M.N. Toksoz, "Detection of Leaks in Subsurface Liners Using Guided Acoustic Waves," ASCE National Conference on Environmental Engineering, July 1-3, 1985, Boston, MA.

Maser, K.R. "Automated Interpretation of Sensor Data for Evaluating In-Situ Conditions", First International Conference on Applications of Artificial Intelligence to Engineering Problems, April 15-18, 1986, Southampton University, U.K.

National Cooperative Highway Research Program (NCHRP), "Durability of Concrete Bridge Decks," NCHRP Synthesis of Highway Practice No. 57, May 1979.

Nii, H.P., and E.A. Feigenbaum, "Rule-Based Understanding of Signals," from D.A. Waterman and F. Hayes Roth (eds.), Pattern Directed Inference Systems (New York): Academic Press, 1978).

Nii, H.P., E.A. Feigenbaum, J. Anton, and A.J. Rockmore, "Signal-to-Symbol Transformation: HASP/SIAP Case Study," The AI Magazine, Spring, 1982.

Shortliffe, W.H., Computer-Based Medical Consultations: MYCIN, Chapter 3, New York: Elsevier/North-Holland 1976.

Smith, Reid, G., "On the Development of Commercial Expert Systems," The AI Magazine, Fall, 1984.

Ulriksen, C.P.F., Application of Impulse Radar to Civil Engineering, Doctoral Thesis, Lund University, Department of Engineering Geology, 1982.

Whittaker, W.L. et al. "First results in Automated Pipe Excavation," Second Conference on Robotics in Construction, Carnegie Mellon University, June, 1985.

QUALITATIVE PHYSICS AND THE PREDICTION OF STRUCTURAL
BEHAVIOR

By John H. Slater,[1] A.M. ASCE

ABSTRACT

   A key need in the application of expert systems to structural engineering design is the ability to predict structural deformation and
stresses in the absence of a quantitative model. The use of approximations and general rules for predicting deflections and moments forms a
necessary part of the preliminary selection and design procedure. If a
program is to aid in the synthesis of new configurations, a part of its
knowledge must include the qualitative aspects of structural behavior.
This paper discusses the application of a rule based approach to the prediction of deflections and moments in indeterminate beam structures. It
is being implemented in an Intelligent Computer Aided Instruction (ICAI)
program to critique proposed deflected shapes and moment diagrams.

INTRODUCTION

   In generating a preliminary design for a structural framework composed of beam members, the engineer must make use of rules of thumb to determine approximate deflections, axial forces and bending moments. These
are then used for initial proportioning of members. While accurate matrix
analysis methods (displacement methods) exist, they require a complete
specification of the structural model for both statically determinate and
indeterminate structures. The minimum data required includes the exact
geometry, loads, boundary conditions, materials and cross sectional properties. This detailed information is typically not available at the onset
of the preliminary design phase. In addition, a matrix approach is an application of deep engineering knowledge which obscures the effects of individual members and their characteristics. It is virtually impossible
for a computer program using a matrix method to determine the effects of
changing a beam's length without rerunning the entire analysis and compar-
ing solutions, since <length> is known only by its effect on certain
stiffness coefficients and load terms. Hence, matrix methods are primarily useful with a generate and test paradigm for structural design.

   Various approximate methods have been developed for structural analysis which map indeterminate structures to geometrically similar determinate structures. These require only the specification of geometry,
loads and boundary conditions together with the application of statics for

---

[1]Assistant Prof., Department of Civil Engineering, MIT, Rm 1-272,
Cambridge, MA   02139

a complete stress analysis [eg. 1]. These mappings introduce sufficient assumptions about the values of moments, forces and shears in columns or girders to render the system determinate, removing the dependence of forces and moments on material stiffness and cross sectional properties. In HI-RISE [2], an expert system for the initial proportioning of plane frames subjected to lateral and gravity loads, mapping approaches are applied iteratively to size members in the framework.

To give a computer program the ability to propose new candidate designs, surface approaches which localize the effects of given parameters and which do not require the solution of a large set of simultaneous equations are needed. The level of accuracy is correspondingly reduced, but the understanding of effect is increased. Ideally, a qualitative solver would be used in conjunction with a quantitative matrix model which would be refined as the design/analysis cycle progresses.

In this paper, a qualitative approach to preliminary structural analysis is presented in the context of a rule based system for the determination of approximate deflected shapes and bending moment diagrams for continuous beam structures. Only qualititative values eg. <large, medium, small, zero> are known for span lengths, loads and cross sectional stiffness EI. The rule based system is initially being developed as an ICAI tool to aid students in sketching deflected shapes and moment diagrams. The rule based approach allows error diagnosis and tutoring at the fundamental level of structural behavior, where absolute values are not the issue. Students will electronically sketch a proposed shape or moment diagram, building it from a set of given component functions -constant, linear, parabolic, etc. - using a menu. The input solution will be verified against the qualitative rules, and an appropriate diagnosis made. The use of the rule based system on its own will be beneficial as a demonstrative tool to aid the student in learning when and how to apply particular strategies. A longer term goal of the project is to use the qualitative analyzer as a tool to guide a generate and test approach to design. This will be especially useful in altering a given structure to meet given requirements such as "increase the positive moment", or "decrease the midspan deflection".

BACKGROUND

A protocol study of structural analysis problem solving technique was undertaken by Cowan [3,4] to determine the methods used by engineers in qualitatively sketching moment diagrams and deflected shapes for continuous beam structures. Subjects were asked to produce approximate solutions to indeterminate structures for which values of loads and span lengths were unknown. There basic classes of problem solver were identified, deflection based solvers, force based solvers and algebraic solvers.

The deflection based solvers rely on the kinematics of deformation to solve the problem. They are able to sketch a deflected shape by first placing inflection points at appropriate positions for each span, determining known points of zero deflection or rotation, and intuitively estimating the sense of curvature. Once the approximate deflected shape is known, they sketch the moment diagram by placing known shapes corresponding to the type of load in each span. The inflection points are used to locate zeros in the moment diagram, since moment = EI x curvature and curvature is zero at points of inflection.

The force based solvers rely primarily on statics to solve the problem. They begin by identifying the sense of reactions and reaction moments, and then place moment diagrams on each span corresponding to known simple span solutions. These are then pushed up or down to satisfy moment equilibrium at the nodes between adjacent spans. Once the moment diagram is determined, the corresponding deflected shape is sketched. This technique is effective for problems of two or three spans, but falls apart for larger problems, and for problems with complicated boundary conditions, since it relies heavily on remembered solutions to determine the reactions initially.

The algebraic solvers try to solve the problem abstractly by expressing moments and deflections algebraically using the appropriate equations. This eliminates the need for explicit values of load and span length, but is insufficient for any but the most simple indeterminate structures.

Cowan identified several successful qualitative strategies used by experienced engineers [3]:

*   Pushing a parabolic shape according to a checklist

*   Using superposition of known solutions

*   Breaking structures into sub-structures

*   Superimposing moment envelopes

*   Releasing rotations or deflections and then rejoining

*   Sketching deflections

*   Determining reaction directions

For a rule based system, certain of these strategies would be quite difficult to implement, since they require rather sophisticated pattern recognition and recall of previous solutions. The initial determination of reaction directions, for example, relies on an intuitive solution of the equilibrium equations for each span simultaneously, which is generally based on a previous known solution. Breaking of structures into sub-structures relies on a recognition of nodal support patterns in discrete parts of the structure, a difficult task to implement for the general case.

An effective approach can be developed by combining a "force" strategy for a given loaded span with a "deflection" strategy to determine how that span affects others. It is relatively straightforward to incorporate known solutions for single spans in the knowledge base of the program, either using rules or functions to be invoked when the appropriate conditions are met. Next, continuity and the concept of rotation and moment "carry over" from one end of a span to the next can be used to determine certain end rotations and curvatures. The concept is much like single pass moment distribution. Finally, equilibrium and/or kinematics can be applied to individual members to refine the solution. An effective feature of the rule approach is that the rules will fire in the order appropriate for a given problem, and the strategy does not need to be preordained. This gives greater flexibility in the type of problem which may be solved, and from the point of view of the student user, observation of which rules are fired when can be used to develop an intuitive sense of problem solving strategy.

A "kinematically complete" solution for the deflected shape of a beam structure is attained when qualitative values for deflection and rotation are established for the ends and center span of each member. Correspondingly, a "statically complete" solution for the moment diagram is attained when qualitative values for moment (or curvature) are established at the ends and center. Given the end and center values, it is possible to draw the deflected shape (using splines, for example) and to draw the appropriate bending moment diagram. In these solutions, the qualitative values <large>, <medium>, <small> and <zero> will be used as quantifiers, along with their negative counterparts.

PROBLEM DEFINITION

The continuous beam structure to be considered is defined by a set of straight members connected by nodes. For ease in description, it is assumed that each member spans from left to right and has two nodes, l_node and r_node. Five node types are recognized, corresponding to typical boundary conditions (bc's):

    fixed: displacement is zero rotation is zero

    pinned: displacement is zero rotation continuous

    pinned hinge: displacement is zero rotation discontinuous

    span hinge: displacement continuous rotation discontinuous

    free: displacement discontinuous rotation discontinuous

This set of nodal boundary conditions allows a potentially large class of members, including cantilevers, links, continuous beams and fixed end beams.

    Member loads are of two basic types,

    uniform: distributed load per unit length

    concentrated: point load

and loads may be directed upwards or downwards. In the case of concentrated loads, they may be positioned at the left end, center or right end of the span.

Rules of thumb and known solutions involving the mathematical combination of terms will be required to determine qualitative values for rotation, deflection, shear and moment. For example, rotation x lever arm equals deflection. If rotation is known to be <large> and the lever arm is known to be <large>, deflection will also be <large>. It is possible to define a qualitative arithmetic that deals with the values <large>, <medium>, <small> and <zero> in an unambiguous way, but the granularity of the resulting operations leads to serious roundoff error, especially for a sequence of operations. A computationally more convenient approach is to associate a relative numerical value with each quantifier and to use these relative values in any needed arithmetic operation, using the values

<large>   1.00
<medium> 0.67
<small>   0.33
<zero>   0.00

and corresponding negative values for <large_negative> etc.

In the beam problem, it is necessary to deal with different types of quantities: deflections, rotations, moments, shears, span lengths and load magnitudes. The qualitative arithmetic which combines these quantities can be made more accurate if <large> is interpreted to mean large with respect to some established reference value. In structural engineering problems, known solutions can be used to establish typical reference values. For example <large> for span length of a beam would be 40 feet, <large> for deflection would be on the order of 1% of the span length, etc. A convenient tactic is to convert input qualitative quantifiers for load magnitude and span length to relative numerical values based on sensible reference values, to use these relative values in determining deflections and curvatures, and finally to convert results back to the equivalent qualitative quantifer prior to presenting them to the user.

SOLUTION STRATEGY

If a structure has only one loaded span, the direction of displacement beneath the centroid of the load is known to be in the direction of the load (Conservation of Energy). This provides a good starting point from which the deflected shape and moment diagram may be deduced. Suppose for example the middle span of a three span continuous beam is loaded uniformly downwards. The center span will deflect downwards, the left end will rotate counter clockwise and the right end clockwise to accomodate this deflection. The actual amount of rotation will vary between zero and the free end rotation of a simple beam, depending on the rotational stiffness of the adjacent spans. For the general loaded span this may be expressed as

$$end\_rotation = (C_1 \times center\_deflection) / length$$

Values for C_1 are bounded by 0.0 for complete rotational fixity and 3.20 for simple ends. A value of $C = 2.5$ was chosen for the typical continuous end after examination of several exact solutions.

Rotations at the near end of a member induce rotations at the far end.

$$far\_end\_rotation = C_2 \times near\_end\_rotation$$

If the far end is not free to displace, the induced rotation factor will vary from $C_2 = 0.0$ if the far end is fixed to $C_2 = 0.50$ if the far end is pinned. For continuous far ends a rotation carry over factor $C_2 = 0.30$ is appropriate in the typical case. A similar carry over factor is used for far end moments induced by near end moments. Special rules are used for cantilever members (statics always satisfied first, then rigid body kinematics) and for members whose far ends are connected to links unable to resist transverse motion.

The shape or order of the moment diagram and deflected shape for each
span may be deduced from the type of loads present, since shear is the
integral of load, moment the integral of shear, and slope the integral of
moment. This information may be used to deduce the location and number of
inflection points, for example.   If the end rotations of an unloaded span
have the same sense and there is no relative displacement, there   must be
one point of inflection.   This in turn implies that moment must be of
opposite sign at each end.   This fact can be used to identify and resolve
conflicts which arise from multiple loaded spans.

At the present time, conflicts are resolved using a superposition
approach. A qualitative solution is obtained for each loaded span and the
relative values for each component in each solution are added.   In the
future a more direct conflict resolution strategy is to be developed which
will take into account relative span lengths, load magnitudes, unbalanced
moment and "distance" from the load. The distance effect follows the
general principle that effects of any given load diminish with an increase
in the number of spans between the point of application of the load and
the point of interest.

IMPLEMENTATION

The qualitative beam analysis program is being implemented using
GEPSE [5] a forward chaining expert system shell written in C that operates on the IBM PC and UNIX mini-computers.   GEPSE uses frames to store
data about objects, and rules or functions to manipulate objects and
attributes.   The forward chaining inference mechanism (appropriate for
this problem) has facilities for meta-rules to change the active rule
base.   GEPSE provides an object network language, ONL, which may be used
to write functions to operate on objects.   A standard library is provided
for attribute extraction, instantiation and logical comparisons.   Attributes may be multiple valued, which provides a convenient mechanism to
explain how a particular value was arrived at - one can simply attach an
"explanation value" each time an attribute is changed.

Rules are of the form

IF <antecedent list>

THEN <consequent list>

Antecedents involve logical comparisons of attributes and objects.

Consequents may involve attribute extraction, instantiation, computation using ONL functions or compiled C language functions and programs, or
running the inference engine recursively on a different set of rules. The
inferencing is goal directed, and rules are typically packaged in "knowledge bundles" or rule bases, which apply to different classes of problem.
For example, in the qualitative beam analysis package bundles of rules are
used to classify members according to end boundary conditions <cantilever,
continuous, etc.>, to apply statics to a given node or member, to determine initial deflections for loaded members, or to propagate effects from
one selectively on members or nodes.

Rules are written using the ONL language with a lisp-like syntax. For example, the rule which sets the left end rotation of a continuous beam from a known center deflection is written in the GEPSE ONL as

```
(make_rule l_theta_from_c_delta
      (and (is? ako ?BEAM member)
           (known? c_delta ?BEAM)
           (is? theta_continuity (get l_node ?BEAM) continuous)
           (unknown? l_theta ?BEAM)
      )
      (do  (set l_theta ?BEAM
             (* -3.0 (/ (get c_delta ?BEAM) (get length ?BEAM))))
           (add l_theta ?BEAM ">l_theta is -c_delta*3/length")
      )
)
```

Translated, this means

IF
>   object BEAM is a member, and the center deflection is known and
>   the left node of the BEAM is continuous and the left rotation is
>   unknown,

THEN
>   set the left rotation to -3/length times the center span
>   deflection and add an explanation for that value.

The final antecedent, (unknown? l_theta ?BEAM) prevents the rule from firing more than once, since the consequent of the rule sets the attribute l_theta on the first firing, making it known.

The overall solution strategy is guided by meta rules as follows:

1) Set object types and check input data for completeness-load data connectivities, etc.

2) Classify nodes for rotation and deflection continuity

3) Classify members as to type

4) Set known values of deflection, rotation, shear ane moment from nodal boundary conditions

5) Deduce possible shear and curvature distributions in members from loading

6) Estimate centerline deflections and end moments of loaded members from known solutions

7) Propagate end effects and apply equilibrium and kinematics to determine displacements from end rotations or shears from moments or moments from shears, etc.

8) Adjudicate conflicts arising from multiplying loaded members by superposition

9) Convert relative values to qualitative ones

10) Display results and explain reasoning by printing the "explanation values"

An example solution for a sample five span beam is shown in Figure 2. The rule firing sequence was

1) determine center span deflection from fixed end solution

2) deduce end rotations from center deflection

3) apply rotation continuity at the ends of the center span

4) apply moment balance at the ends of the center span

5) carry over moments to far ends 6) repeat steps 4 & 5 for exterior spans

Case A indicates the problem as posed.   There are five roughly equal spans, all of span length <large>, and a uniform load on the center span of magnitude <large>.   After an initial classification of members, and setting of displacements to zero at the ends of members adjacent to supports and moment to zero at the extreme ends, control passes to a set of rules to determine deflections and moments for the loaded span.

In Case B, moments l_moment, c_moment and r_moment have been deduced for the center span, and an initial deflection value for c delta has been determined.   The user is notified that all loaded spans have been solved, and an explanation "Fixed end solution" is attached. to each value.

In Case C, control has passed to the rule base for distributing moments and rotations.   Continuity of rotation has been applied at either end of the loaded span, moment balance has been applied, and a carry over of 30% of the end moment and rotation to the 4th span has occurred. Again, each new value is tagged with an explanation.

In Case D, a complete solution has been attained, solely by carry over and continuity. The corresponding deflected shape is shown in Case E. The center span deflection is roughly twice the first interior span deflections, which are in turn roughly 2.5 times the exterior span deflec- tions.   End rotations over the first interior supports are 30% of rota- tions in the center span, and the exterior support rotations roughly 10% of the center span amount.   These results are quite reasonable, as they should be since a limited cycle moment distribution with nodified carry over factors was accomplished by the rule based system.

FUTURE WORK

The work presented so far is in the developmental stage. The rule approach is being extended to systems for which the solution is more detailed than the beam continuous over supports.   In particular, a better "understanding" of behavior will be built into the rules, if possible, so that they need not rely quite so heavily on the carryover factor approach.

The advantages of the package from a teaching point of view are manifest.   The program may be used to demonstrate solution approach for

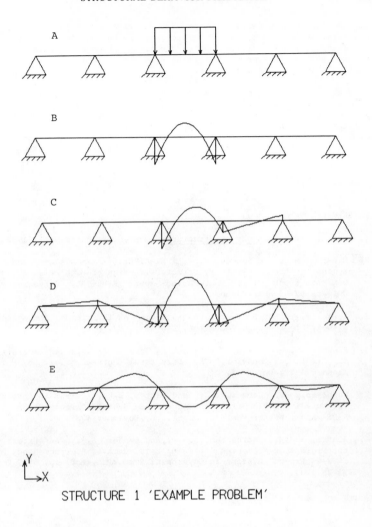

STRUCTURE 1 'EXAMPLE PROBLEM'

Figure 1

various problems in indeterminate beam analysis, and importantly, EXPLAIN its reasons for doing what it is doing at the time that it is being done. The logic in the qualitative solver is placed in two areas, in the meta-rules which govern the basic steps of the problem solution, which are essentially the same from problem to problem, and in the antecedents to the rules for distribution, balance, initial solution etc. This logic will be particularly effective when the tutoring module is added to judge deflected shapes and moment diagrams input by the student. The strategy to be employed will involve backward chaining through the rules fired by the qualitative solver in obtaining its own solution, and identifying contradictions in the student input.

A similar tutoring approach is taken in MACAVITY[6], a rule based package which diagnoses and explains errors in equilibrium equations written by the students for determinate beam problems. Tutoring and explanation capabilities are an important addition to the capabilities of ICAI for educating engineers about problem solving technique, and have po-tential extension to the engineering world of the professional as well.

REFERENCES

1. Hsieh, Y.Y., Theory of Structures, 2nd ed. Prentice Hall, 1982.

2. Maher, M.L. and Fenves, S.J., HI-RISE: A Knowledge-Based Expert System for the Preliminary Structural Design of High Rise Buildings, Department of Civil Engineering Report R-85-146, Carnegie Mellon University, January 1985.

3. Cowan, J., "How Engineers Understand: An Experiment for Author and Reader," Engineering Education, January 1983, pp. 301-304.

4. Brohn, D.M. and Cowan, J., "Teaching Towards An Improved Understanding of Structural Behavior," The Structural Engineer, Vol. 55, No. 1, January 1977, pp. 9-17.

5. Chehayeb, F., Connor, J.J. and Slater, J.H., "GEPSE-An Environment for Building Engineering Knowledge Based Systems," Proceedings of the ASME Annual Winter Meeting, Miami, FL, November 1985.

6. Slater, J.H., Petrossian, R.B.P. and Sunder, S.S., "An Expert Tutor for Rigid Body Mechanics: ATHENA CATS -MACAVITY", Proceedings of the IEEE/CS Expert Systems in Government Symposium, McClean, VA, October 23-25 1985.

SUBJECT INDEX
Page number refers to first page of paper.

AUTHOR INDEX
Page number refers to first page of paper.